T0176352

# Dr. Rip's ESSENTIAL
# Beach Book

# Dr. Rip's ESSENTIAL

# Beach Book

**Everything You Need to Know
About Surf, Sand, and Safety**

## ROB BRANDER

Columbia University Press
New York

Columbia University Press
Publishers Since 1893
New York    Chichester, West Sussex
cup.columbia.edu

Library of Congress Cataloging-in-Publication Data
Names: Brander, Rob, author.
Title: Dr. Rip's essential beach book : everything you need to know about surf, sand, and safety / Rob Brander.
Description: New York : Columbia University Press, 2024. | Includes index.
Identifiers: LCCN 2024001453 (print) | LCCN 2024001454 (ebook) | ISBN 9780231217392 (hardback) | ISBN 9780231217408 (trade paperback) | ISBN 9780231217415 (ebook)
Subjects: LCSH: Beaches. | Beachgoers—Safety measures. | Geomorphology. | Coastal zone management.
Classification: LCC GB451.2 .B73 2024 (print) | LCC GB451.2 (ebook) | DDC 551.45/7—dc23/eng/20240214
LC record available at https://lccn.loc.gov/2024001453
LC ebook record available at https://lccn.loc.gov/2024001454

Printed in the United States of America

Design Josephine Pajor-Markus
Cover design Madeleine Kane
Cover images Front: The Pass at Byron Bay, New South Wales. Shutterstock / Darren Tierney. Back: White sand beach at Jervis Bay, New South Wales. iStock / Veni

*To my Mum and Dad who gave me
their love of the sea (and the beach!)*

# Contents

# About this book

This book is for anyone who loves the beach. I'm thinking that's a lot of people. The allure of beaches runs deep regardless of whether we visit them once a year on holiday or every day for a walk, swim or surf. People travel thousands of miles and spend a fortune to visit beaches, so if you happen to live close to one, keep reminding yourself how lucky you are, because they are exceptionally beautiful, special and fun places.

My own love of the beach is wrapped up in memories of childhood vacations to Cape Cod on the east coast of the United States. After days of driving in the car with my parents from the suburbs of Toronto, Canada, my anticipation levels would slowly build as the first signs for the Cape appeared on the Massachusetts Turnpike. They increased rapidly as the sides of the roads became sandier and reached a crescendo of excitement as we finally crested the last hill and were greeted simultaneously by the glorious reflection of the sun off the water, the smell of the sea and the sounds of the beach. It was the best feeling in the world … and still is.

Beaches mean different things to different people, mostly because beaches can be so, well, different. The motivation for writing *Dr Rip's Essential Beach Book* was not to provide a guide to the best beaches in the world or to capture our romance with them, but rather to try and explain what makes your favourite beaches tick. It's really a book about coastal geomorphology.

When people ask me what I do for a living and I tell them 'I'm a coastal geomorphologist' I get some odd responses, usually of the thousand-yard stare variety. What most people don't know is that coastal geomorphology is the study of how beaches work and, although they may not realise it, anyone who goes to the beach has a vested interest in that. It tells us where the pile of golden sand we lay our towels on came from, why waves break and give us good surf, how the tides that wash up new treasures each day work, what causes the deadly rip currents that take us out to sea, why there are so many amazing different types of beaches and why some of them are disappearing. And much, much more.

The idea for this book really began in 2001 when I started giving community talks called 'The Science of the Surf' at my surf life saving club at Tamarama Beach in Sydney's Eastern Suburbs. Some of my fellow lifesavers thought it would be a good idea to teach the locals and tourists about how waves, rips and beaches worked. Word spread, the talks became more popular and people kept coming up to me saying, 'I've lived on the coast all my life and didn't know any of this stuff!' I got the impression that surfers, swimmers, parents and tourists alike all had an innate desire to know a little bit more about their favourite beaches. This is the written version of 'The Science of the Surf'.

If the beaches you walked upon could talk, they could tell you a fascinating story and that's what I've tried to capture in the next six chapters. Chapter 1 begins with an explanation of how our beaches, down to their little sand grains, were formed and why we find them where we do. Chapters 2 and 3 describe the waves, tides and storms that impact our beaches. Breaking waves not only create good surfing but also some dangerous currents, particularly rip currents, which are the focus of chapter 4. Beaches in all their glory, from the various features we find on them, to the different types that exist, to the impacts we have on them are all described in chapter 5. Finally, chapter 6 provides a brief guide to some of the physical and biological hazards that you might find on beaches and how you can avoid them.

It doesn't matter whether you just read the bits that interest you or read the book from cover to cover, I just hope that at some point you'll find something that makes you think, 'Wow, that's really neat' and inspires you to keep reading for more. Although the book is about beach science, I've tried to explain how things work in a way that is hopefully easy to understand and fun. The great thing about beaches is that they all follow the same fundamental principles, whether it's Bondi Beach in Australia, Malibu Beach in California or Kuta Beach in Bali, so I have tried to use examples from beaches around the world.

There is also a beach safety theme throughout because far too many people drown and get seriously injured on beaches and that just shouldn't happen. A little knowledge can go a long way to making your visit to the beach that much safer. Finally, throughout the book can be found sections that describe quirky things about beaches and 'Using the Beach' sections for those who want to expand their beach experience a bit more.

This is the second edition of the book and in the thirteen years since the first edition was published I am pleased to say that I've yet to see a copy in a second-hand bookshop! It's possibly because there aren't many second-hand bookshops left, but I'd like to think this book is a 'keeper' because it's packed with useful information that can also help people. While the basic science of beaches hasn't really changed, our understanding of beaches and beach safety is always improving, including my own. I've updated this edition to include some of the latest science, particularly in relation to sea level rise and storms, which both contribute to beach erosion and are impacted by climate change. Funnily enough I was never really happy with how I dealt with rip currents and beach safety so I've greatly improved chapters 4 and 6. I've updated some terminology and provided some more recent examples of things throughout. The book also has a brand new look, including a dedicated picture section. I've also included a few more personal stories, mostly because I can, but people said they liked them!

I've also tried to mention some coastal geomorphologists who have not only helped me, but have also made a big contribution to our understanding of beaches. Scientists often don't get the credit they deserve and we wouldn't know much about beaches without their efforts. Coastal geomorphologists are also not stereotypical, geeky little scientists. Most spend a lot of time in the surf and have some amazing stories to tell. You should really try talking to one someday.

# Dr. Rip's
## ESSENTIAL
# Beach
# Book

# 1

# The sands of time

## Beaches past, present and future

The beach. There's a lot wrapped up in those two words. Close your eyes and all sorts of images flash through your mind. A strip of golden sand, the sounds of crashing surf, the endless blue ocean stretching to the horizon, clean lines of swell forming perfect right-handers, the smell of sunscreen, a hammock swinging underneath palm trees … maybe even a screaming jetliner above your head (see Pic. 1 in the picture section)! The point is we all have our ideal image of a beach and it's really not that hard to find. We are literally surrounded by beaches and most of them are simply fantastic.

We may all know a beach when we see one, but few of us understand them. The reality is that every beach is unique, both in its charms and in its dangers. Each year, tens of thousands of people are rescued from the surf, while far too many others are not so lucky. Understanding how a beach works is important to everyone who visits one and there's no better place to start learning this than on the beach itself, with the sand between your toes …

Piha Beach stretches away into a horizon misted in sea-spray along the rugged west coast of Aotearoa New Zealand. The beach, a popular day trip for residents of nearby Auckland, is famous for its dramatic scenery, nasty rip currents and remarkable inky-coloured sand. Several years ago I was walking along the beach on a beautiful day and decided to keep my thongs ('jandals' to Kiwis – no, I have no idea either!) on because black sand has a tendency to become hot when it's sunny. Sure enough the soles of my feet soon started to feel the pleasant warmth of the sand penetrating through the heat shield of my trusty thongs. Unfortunately this was soon replaced by a distinct softening sensation that quickly turned into searing heat and the feeling that that my feet were on fire. Looking down at the melting rubber curling around my toes I was overcome by a sudden urge to sprint to the ocean.

Reaching the cool water I stood amid the hissing sounds and clouds of steam emanating from my feet, feeling the type of relief usually associated with reaching nirvana. Looking around, I realised that I was one of the lucky ones. Hundreds of beachgoers were stranded along the beach trying to escape the black inferno by clinging to small islands of towels, staring helplessly at the distant safety of the ocean and car park. Some were trapped balancing on one leg, seeking relief from the meagre shade provided by clumps of seaweed and the odd patch of dune grass, while contemplating their next agonizingly painful move. I'm not sure how many people were feeling existential and thinking, 'Why is the sand black?' – not many I suspect – but it would have been a pertinent question to ask. An even better question would have been 'Where did this sand come from in the first place?'

The moral of this story is that the sand that makes up a beach is more important than you might think. People flock to beaches for all sorts of reasons, otherwise millions of people wouldn't visit them every year. And it's not just people who end up at the beach. Seaweed, flotsam and jetsam, whales and the odd cargo ship all have a tendency to wash up from time to time, but ultimately beaches are the final resting place for billions and billions of tiny sand grains.

Next time you go to the beach, scoop up a handful of sand and look at it. In fact, try doing this for every beach you visit for the rest of your life. What do you see? Does the sand always look the same? What is it made of? As much as we love them, beaches are very much something we take for granted and despite billions of tourism dollars wrapped up in these little parcels of sand, how the sand got there in the first place remains a mystery to most of us.

As with any journey there is always a story involved, and in the case of sand, it happens to be a fascinating one. To understand how our favourite beaches were formed, we need to go back a long way in time and as Professor Bruce Thom, one of Australia's pioneering coastal geomorphologists, once told me, when it comes to the beach, 'It's all about the rocks' (Pic. 2).

## It's all about the rocks

### The colours of beach sand

Believe it or not, beaches are a dump. Fortunately they happen to be a natural one for all the materials broken down by thousands and sometimes millions of years of erosion of the earth's landscapes. If you look closely at the handful of sand you've picked up, you'll see an awful lot of individual sand grains. Are all of them exactly the same colour and size? Usually you'll find a mixture of sand grains that are often from different rock sources and have a range of sizes. My university students may find analysing sand a bit boring, but they always perk up when they have a look at a sand sample under a microscope and see a kaleidoscope of colours and shapes (Pic. 3). However, on most sand beaches, most of your handful of sand is made of quartz minerals.

There's a good reason for this. The most common type of rock on the earth's surface is granite, which just happens to contain a lot of minerals, including quartz. When granite rocks are broken down and eroded, many of these minerals are released and end up being carried by rivers, wind and glaciers on a long journey that may or may not end up at a beach. Quartz is rather special because it is exceptionally hard and resilient and while other minerals can break down and almost completely disappear, quartz doesn't. So by process and elimination, quartz ends up on a lot of beaches around the world.

However, as any budding mineralogist will know, a pure quartz crystal is a nice clear, whitish colour. This would mean that most beaches should be completely white in colour, and there are definitely some stunning white sand beaches but they are the exception rather than the rule. In reality, the range of colours of beach sand around the world is immense and can vary from shades of grey and brown to red, black and even green.

I know this because I collect sand. I'm not sure why I started doing this other than it seemed like a good thing to do at the time. I'd collect sand from beaches I visited and put them in my mother's old spice jars, dumping out encrusted cardamom seeds that hadn't been used in decades, and then

# THE BEACH WITH THE WHITEST SAND IN THE WORLD

I have a postcard from the south coast of New South Wales that says 'Hyams Beach: the whitest sand in the world'. Sounds impressive. But how do they know for sure? Siesta Key Beach in Florida apparently has the 'official whitest sand in the world'. Then there is Whitehaven Beach in Queensland's Whitsunday Islands, which has sand so white you can develop snow blindness. The truth is there are white sand beaches around the world and probably no single beach can really claim to have the whitest sand of all. After all, how white is white? There are two ways to get pure white beaches. First, really, really, really old sand that has been eroded down through the aeons to its quartz crystal core will be white in colour. Second, many tropical beaches contain broken shell and coral fragments, made of calcium carbonate (limestone), which is whitish in colour and can become even whiter by bleaching in the tropical sun. Unfortunately, there is a downside to beautiful white beaches - the pure quartz is used in making glass, which involves mining the sand. For such a beautiful environment, this is a destructive operation.

# THE CURSE OF THE BLACK SAND BEACH

For many tourists visiting Hawai'i, black sand beaches are quite a novelty and a little jar of the black stuff makes a nice souvenir. What a shame it comes with a curse. Pele, the volcano goddess of Hawai'i, gets upset when people take samples of her lava rocks and black sand away and inflicts a curse of bad luck on them for the rest of their lives. And it's not just minor nuisance type of bad luck, but the cataclysmic life-altering kind! The curse must have something to it as plenty of Hawaiian visitor centres display packages of sand mailed back with profuse apologies. However, dig deeper into the story and there is nothing in Hawaiian legend that mentions bad luck associated with rocks and sand. Instead it seems that a modern National Park ranger got tired of people illegally removing things from the National Parks. The story has grown to the point where the 'curse' now applies to many other black sand beaches around the world. I must admit that Hawaiian sand makes up some of the centrepieces of my personal collection, but fortunately I was aware of the curse and managed to circumvent it by asking unsuspecting relatives and friends to get the sand for me. Did I tell them about the curse in advance? No. Did they experience bad luck afterwards? Yes!

place the sand jars on a shelf in my room. Over time the collection grew and my friends would bring back sand from their trips to beaches around the world. People would always laugh about my hobby until they saw the collection and were immediately impressed because no two sand samples looked exactly the same. The number of sand samples eventually grew to several hundred, and then I moved to Australia where Australian Customs and Quarantine effectively put a stop to my collection! The overseas ones anyway. But Australian beach sands have a lot to offer and I have recently become a fan of beach pebbles, which can make a nice garden feature, but I digress.

So why are there so many different colours of beach sand around the world? Two reasons. First, the age of the sand is important because older quartz sand grains have had more time to roll and slide around, which has scraped away all the surface material around them exposing their pure crystal structure. This makes them and the beach lighter in colour. Younger quartz sand grains, more recently broken off from their source rocks, will still have a darker coating of material stuck to them and the beach will also be a darker colour. Sometimes the coating around sand grains may contain iron, which will essentially rust when exposed to the elements, and can create some beautiful hues of brown, yellow, orange, pink and red sands. The eroding sand cliffs behind Rainbow Beach in Queensland, Australia, are said to have sands of 74 different colours resulting from various combinations of minerals, iron oxides and dyes leached from local vegetation.

The second reason is that there are the other types of minerals found on beaches. Black sand, the bane of many Kiwis, and sometimes supposedly the bearer of bad luck, is a great example of a heavy mineral. Heavy minerals such as magnetite, zircon and rutile are derived from different types of minerals found in granite and other volcanic rocks such as basalt. They are called 'heavy' minerals because they are physically heavier than quartz. They also happen to be darker in colour. Pick up a handful of quartz sand and an equal handful of black sand and you'll be surprised how much heavier the black sand is.

Being heavy, black sand grains on a beach tend to wriggle their way down through the lighter quartz sand into concentrated layers. Sometimes these layers are large enough to be mined commercially as the heavy minerals are quite valuable for industrial uses. Often after a big storm, you may notice black streaks across the beach. Don't worry, it's not oil, it's just some of the black sand layers that have been exposed after the quartz sand has been washed away. Not surprisingly, black sand beaches are more common in areas of volcanic activity such as Hawai'i, New Zealand, the Canary Islands and Iceland (Pic. 4). Hawai'i is also famous for green sand beaches, which often look a bit mouldy at first glance, but are actually made up of olivine minerals formed when lava cools quickly, such as when it flows directly into the ocean. If you look closely on beaches in the Outer Banks of North Carolina or Pfeiffer Beach on California's Big Sur coast after storms, you'll even find stretches of beautiful purple sand caused by another heavy mineral: garnet.

*The eroding sand cliffs behind Rainbow Beach in Queensland, Australia, are said to have sands of 74 different colours ...*

## Are all beaches made up of sand?

Despite all this emphasis on sand, what beaches are really made up of are *sediments*. Sediments are any loose bits of material that have first been eroded from rocks and then moved by water, wind and even ice to be dumped somewhere else, in this case the beach. Some of your beach sediment may be very fine like powder, some might be coarse like small pebbles and some may be made up of cobbles. Some may even be biological in origin, coming from broken bits of shells or coral (Pic. 5).

How big the sediments on your beach are depends on how far they've moved. When rocks are eroded, they start off as chunky and angular fragments. The further they travel, the more they get worn down, becoming smaller and rounder. So the size of sediments on a beach is a clue to how far they've travelled. The smaller they are, like sand, the longer the journey they've taken to get to the beach. If the sand grains are exceptionally round, they will also be noisy.

Noisy? Have you ever walked along a soft sandy beach and with every footstep, the sand squeaks or barks back at you? This doesn't happen on all beaches because you need to have sand grains that are well rounded and almost all the same size. This makes them slide easily past each other and the friction creates sound. They don't call Squeaky Beach in Wilsons Promontory National Park, Victoria, Australia, 'squeaky' for no reason!

Nothing, however, compares to the beach at Padangbai in Bali, the jumping-off point for ferries to the neighbouring Indonesian island of Lombok. The enclosed bay is super-saturated in sticky calcium carbonate, which sticks to sand grains like glue. As the sand grains roll around, more stuff

## ROCKPOOL RAMBLING | Many beaches have

rock platforms and reefs either next to them or poking out in the
middle that are usually submerged at high tide, but exposed for all to
see and explore at low tide. Certain types of marine flora and fauna
are found only in these environments, making rocky intertidal zones
wonderful places of discovery for adults and children alike. Marine
animals that you rarely see, such as octopus, moray eels and baby
sharks, can sometimes get trapped in rockpools during the falling
tide and hang around on display until they disappear again during the
next high tide. Rockpool rambling is fascinating, fun and educational
at the same time.

Rocky intertidal shores offer a range of marine life such as
anemones, starfish and brittle stars, myriads of little fish, nudibranchs,
crabs of different shapes and sizes, as well as corals, algae and marine
plants. The best way to observe how these small communities exist
is to take your time. Life in rockpools can be pretty slow and you may
miss a lot of subtle activity if you rush by. Exploring at night with a
flashlight can be even more rewarding as many marine animals are
nocturnal and more active. A great idea is to go on some of the many
free guided walks provided by various coastal parks, local governments
and community groups, usually during the summer months.

Rocky shores can be potentially dangerous places as they
are inherently slippery and are more exposed to larger waves
than beaches. It's always a good idea to wear sturdy footwear
(I recommend Dunlop Volley tennis shoes) that can grip wet surfaces.
Also make sure you only go at low tide, never when there are big
waves, and stay well back from the waterline. Basically, just use your
common sense. Also remember that some marine animals, although
shy, can either sting, bite or even contain poisonous spikes. The
golden rule with exploring rockpools is 'Look, but don't touch!' Even
if they are completely harmless, marine plants and animals are very
sensitive, so please try not to disturb them in any way.

sticks to it and they get bigger and rounder, forming tiny pellets. It's sort of like making a snowman. Walking across the loose sand beach is crazy. With each step your feet sink in past your ankles, the sand starts barking at you, and the tickling sensation is diabolical. It's probably the only beach in the world where everyone is scampering around giggling to themselves!

Not only can the type of sediment on a beach provide both pleasure (think lovely soft squeaking sand) and pain (think hot black sand), it is also one of the main factors in determining why beaches can look so different. Based on the type of sediment found on a beach, there are really four broad categories of beaches found around the world: sandy, gravel, muddy and carbonate.

## Sandy beaches

It should come as no surprise that sandy beaches are made up mostly of sand. They are also the focal point of this book because they are the ones we interact with the most, are comfortable to lie on and feel good between our toes. While sand beaches occur around the world they won't form unless there's a lot of sand around in the first place. With over 11 000 beaches to choose from, it's no surprise that Australia is called the Lucky Country, but it does help to have some of the oldest rocks on earth, so there has been plenty of time for sand to make its way to the beach. On the other hand, the east coast of the United States isn't as old, but is almost one big, long sand beach from Cape Cod to Florida because of massive amounts of sand left behind from glaciers and from ancient rivers that carried sand out of the eroding Appalachian Mountains.

Sand is mostly brought to the coast by large rivers, glaciers and eroding cliffs. Or it can be transported along the coast from distant sources thanks to something called longshore drift, which is like a slow river of sand moving along a beach. Some of the sand reaches a beach and stays there, but a huge amount also sits underwater, lying on the relatively shallow continental shelves that fringe most of the world's coastlines. Some of this sand can eventually make its way to the beach through the combined efforts of waves, currents, tides and wind. But it takes time. On beaches backed by rapidly eroding cliffs, it's easy to see where the sand comes from as the sand on the beach matches that in the cliff. In other cases, such as the story of Bondi Beach, the sand took millions of years to reach the beach.

## Gravel beaches

Gravel beaches, or shingle beaches as they are known in the United Kingdom, are made up of boulders, cobbles, pebbles and other coarse sediments. They never seem to make it into the tourist brochures because they are difficult to walk over, hurt to lie on and have a habit of launching cobbles the size of grapefruits through the air like missiles when large waves break on them. This does not immediately lend itself to a relaxing day at the beach! Nevertheless, gravel beaches are very common around the world and, while not always the perfect option for a beach holiday, they can be dramatic and beautiful. Just remember to bring a helmet.

As gravels are big and heavy, they don't tend to move very far, which means that gravel beaches are generally made of sediments that have been derived locally. For example, if a beach is surrounded by cliffs full of loose gravels, as the cliff erodes, the gravels fall out, get arranged a little by the waves and *voilà*, you've got a gravel beach. Many gravels found in patches on beaches that are otherwise sandy, such as Old Bar and Coalcliff in New South Wales, are the remnants of ancient river beds that have been exposed through recent beach, dune and cliff erosion. Gravel beaches are more common on coasts in colder climates that have been glaciated and those adjacent to mountain ranges. Plenty are found in New Zealand, the United Kingdom and Europe, as well as the Great Lakes in North America. Perhaps the most famous is Chesil Beach in southern England, which stretches for an impressive 30 kilometres and is a UNESCO World Heritage Site.

## Muddy beaches

Muddy beaches are generally not on the top of many people's list of holiday destinations as they are not great for swimming and tend to smell, but they are very common and are host to a phenomenal range of flora and fauna. 'Mud' is the not-so-polite term used to describe extremely small and fine sediments like silt and clay, which come from the breakdown of softer rock minerals found in granite rocks. Muddy sediments also contain a small amount of organic matter, ranging from bits of vegetation to worm droppings (hence the smell). Silts and clays form more rapidly in hot and humid climates and most mud is brought to the coast by big rivers, which are also more common at lower latitudes. This helps explain why 75 per cent of tropical coastlines have muddy beaches and often nasty things like crocodiles, alligators and sand flies!

*... 75 per cent of tropical coastlines have muddy beaches and often nasty things like crocodiles, alligators and sand flies!*

# THE STORY OF BONDI BEACH

**A** must-see for most international tourists, Bondi Beach is Australia's most iconic beach and is National Heritage Listed. The land has been home to the Bidjigal, Birrabirragal and Gadigal people for millennia, and the story of how it formed is millions of years old.

Bondi lies tucked between two sandstone headlands and the colour of the sand on the beach is very similar to the colour of the sandstone rocks (Pic. 2). So the beach sand comes from the rocks, right? Well, yes, but not the way you might think. About 300 million years ago Australia was part of a supercontinent known as Gondwana, which also consisted of South America, Africa, India, New Zealand and Antarctica. The Australian landscape was already being shaped at this time with mountains to the north, south and west of present-day Sydney leaving a bowl-shaped basin in the middle. About 230 million years ago, while dinosaurs ran amok, this basin started to fill up with layers of sediments carried by massive river systems. Some of the sand layers turned into sandstone rock, most notably a 300-metre-thick layer that people walk along today called the Hawkesbury sandstone.

As the supercontinent slowly started to break up, Australia gradually moved northwards, with New Zealand breaking away about 120 million years ago.

The divorce was slow, taking about 40 million years as the Tasman Sea opened up to the north like a zipper between the two land masses. This left the brand-new east coast of Australia with a steep and narrow continental shelf exposed to large waves that started to slowly erode the Hawkesbury sandstone cliffs, leaving behind a rock platform and a very indented coastline. The opening of the Tasman also triggered more river erosion on the Australian continent resulting in huge amounts of sand being dumped on the continental shelf.

During the last 2 million years, the sea level has been going up and down like a yo-yo. The last ice age was 18 000 years ago and the sea level was 120 metres lower than it is today. Present-day Bondi wasn't even a beach, but a small valley covered in native vegetation, because the beaches were about 25 kilometres to the east. As the earth warmed up, the ice melted and the sea level rose rapidly, stopping about 6500 years ago to the level it is today. As it rose, the sand lying on the continental shelf was bulldozed landward, ultimately filling up the coastal indentations and old river valleys when it stopped. And that is how Bondi formed. The sand you see today is pretty much the same sand that's been sitting there for about 6500 years, but the sand grains themselves are millions of years old and probably came from ancient inland Australia. This might not make for a hot topic of conversation among the sun worshippers on the beach, but it's a cool story.

Muddy beaches are also very common on coastlines that have very gentle wave conditions. For this reason, estuaries, deltas, lakes and lagoons often have muddy beaches and a totally different ecology to sandy beaches, with mangroves, seagrasses and shellfish in abundance. However, if you add water to silt and clay, things can get rather messy and you can literally get mired in the muck, so it pays to be careful while walking along muddy beaches. They are found all over the world, but are particularly abundant along the northeast coast of Australia, monsoonal Asian coastlines, the Brazilian coast north of the Amazon River and the Mississippi delta region in the Gulf of Mexico.

## Carbonate ('coral') beaches

Beaches made out of carbon? Is this possible? Well, sort of. 'Carbonate' is the term used to describe sediments produced by biological processes and living organisms that are made of calcium carbonate, more commonly known as limestone. For example, the shells of most marine animals and the skeletons of corals and coralline algae are really calcium carbonate. As these living organisms grow and eventually die, a lot of this material gets broken down into smaller sizes and is carried to the beach by waves and currents. Even sandy beaches will usually have a small amount of broken-up shells mixed in with the sand.

Where are the best carbonate beaches found? Beaches behind fringing coral reefs, all those stunning little coral cays, like Heron Island in Queensland and all of the islands in the Maldives are great examples of beaches that are almost all 100 per cent carbonate material that has been eroded from nearby coral reefs. Carbonate beaches are also found along rocky coasts wherever sea grass meadows, tidal flats and continental shelf environments provide perfect conditions for the growth of shellfish. Shell Beach in Western Australia is one of only two beaches in the world that is made up almost entirely of the same kind of shell, hence the name.

Whether they know it or not, most people are already familiar with carbonate beaches. They are the lovely white tropical island beaches fringed by turquoise-coloured water shown in all the travel magazines, TV commercials and movies. If you're lucky enough to end up on one of these paradise-like beaches, no doubt you'll want to run barefoot into the inviting water. Well, let's hope you have good travel health insurance! Many coral cay beaches are in fact made up of broken bits of coral that can be hard, chunky, extremely sharp and very painful to walk over (Pic. 5). So if you are planning to head off to some magical coral reef-island beach in Tahiti, do some research before you go and definitely bring along some beach shoes!

## Beaches: stuck between a rock and a hard place

It's easy to group beaches into categories on paper, but many are mixtures of the ones described above. Regardless of the type of sediment found on a beach, in order for a beach to form there has to be a lot of sediment around in the first place. There also needs to be a place for these sediments to sit. In other words, they need to get stuck somewhere. The shape of the coastline becomes really important as all those little nooks and crannies and long straight stretches become potential resting places for large piles of sand (Pic. 2). What controls the shape of the coast? Once again it's all about the rocks.

The shape of any coastline is really just a geologic imprint determined by the rock types and stratigraphy that are there, which will have a range of different structures, hardness and shapes.

The variety of coastal geology around the world is incredible, not only between continents and countries, but even within the same region. Not surprisingly, the variety of beaches can be just as variable. If you don't believe this, try driving around the beautiful coastline of New Zealand or the United Kingdom sometime. The range of beaches boggles the imagination and it's mostly because of the varied geology. On the other hand, vast stretches of the Australian east coast share similar geological characteristics and beaches from Byron Bay all the way south to Eden, a distance of over a thousand kilometres, tend to look much the same. So it's the rocks that provide the sediments and a resting place for them, but the story of beach formation is still not complete. The sediments still need to get to the beach somehow and that really depends on what the level of the sea is doing.

## The importance of sea level

With all the recent attention about global warming, you'd think that sea level rise is a bad thing. It may well be, but if it wasn't for sea level rise, our beaches wouldn't be where they are today. Once sand is delivered to the coast it is then brought to the beach, thanks to the work of waves, tides, currents and wind. However, exactly where these processes occur depends on the level of the sea at the time and sea level has been anything *but* stable for the last 2 million years. It's been rising and falling like a yo-yo because of repeated periods of global warming and cooling.

It's a simple concept. When the earth experiences an ice age, a lot of the water in the ocean turns into ice. With less water in the ocean, the sea level falls. When it's warm, the ice melts, water flows back into the ocean and the sea level rises. There have been times when the sea level has been much higher than it is today. We know this because old beach and sand dune deposits can be found well inland at many places, including Sydney's densely populated Eastern Suburbs.

There have also been times when the sea level has been as much as 150 metres lower than it is today. Think about that for a second. None of the beaches we know today would have existed. Most of them would have been vegetated countryside and a long way from the coast! Land bridges between continents and islands were exposed and you could have walked between Papua New Guinea and Australia, or from Melbourne to Tasmania, if you felt like it. Obviously a lot of people did, as that's how humans are thought to have migrated around the world.

*Sea level also has an amazing control on the rivers of the world, and it's rivers that carry sand to lakes and oceans.*

Sea level also has an amazing control on the rivers of the world, and it's rivers that carry sand to lakes and oceans. When the sea level falls, a river is sort of left hanging and responds by eroding its channel so it can again flow happily into the ocean. This causes a huge delivery of sediment to the coast, causing the coastline and beaches to start shifting seaward. On the other hand, when sea levels rise, rivers respond by clogging up and dumping a lot of the sediment they are carrying inland. Therefore, not as much reaches the coast and the coastline starts to erode and shift inland.

What has this all got to do with beaches we see today? After the last ice age 18 000 years ago, the sea level rose very rapidly as the earth warmed up and the ice melted. The coastline and all the sand that was lying around would have been bulldozed in a landward direction. When it was all over and the sea level stopped rising about 6500 years ago, the ocean had flooded many old deep river valleys, creating estuaries, and filled up smaller valleys and coastal indentations with sand, creating beaches. In other words, many beaches simply got stuck between a rock and a hard place. What most of us don't know is that compared to where the sea level has been over the last 2 million years, the present-day sea level and location of our beaches is quite unusual – we are living in a unique time.

## Here today, gone tomorrow: the future of our beaches?

In geologic terms, 6500 years is not a long time and many coastlines are still trying to adjust to the last rapid rise in sea level. Unfortunately a lot of this adjustment involves erosion, but it is important to remember that erosion is a natural process. In recent years, there has been a huge increase in the public and media awareness of climate change, global warming and sea level rise. This is a good thing because we should all be aware of the impacts that our activities and lifestyles can have on the planet.

However, even scientists can't agree on what is going to happen to the sea level in the next hundred years. It's going to rise, but by how much, no one knows for sure. It could be 20 centimetres, 50 centimetres, 1 metre or as much as 6 metres. It's a complex issue that is beyond the scope of this book. However, there are a few misconceptions about sea level, such as cities becoming swamped and beaches disappearing, that are not necessarily true. Sea level rise doesn't happen overnight. It's a slow process that occurs over decades and centuries and while we can't stop it, we have time to try and mitigate the impacts, mostly by using coastal engineering practices as described in chapter 5.

As for our beaches disappearing, there's good news and bad news. Imagine looking down from a headland at a long, natural beach and sand dune system that is completely natural and undeveloped. Now assume that the sea level will rise by one metre over the next hundred years, which by all accounts, would be pretty bad. If you were somehow able to return to the same headland a hundred years in the future expecting to see the beach eroded away, you'd be in for a shock because everything would look pretty much the same! The only difference would be that the beaches and dunes would have shifted landward thanks to the action of waves, tides and wind. So many beaches won't disappear, they'll just move. That's the good news.

The bad news is that not all beaches are undeveloped. Imagine a heavily urbanised beach where the sand dunes and area behind have been built over by houses and roads. If the same sea level rise occurred, the natural processes would try to shift the beach landward, but wouldn't be able to because of all the development infrastructure. The beach will therefore have nowhere to go and will be eroded by the action of waves and currents. In these situations, sea level rise becomes a big problem, and there are an awful lot of places where we've built structures much too close to the beach, or in inappropriate coastal locations, without an adequate understanding of coastal processes and behaviour.

Measurements have told us that sea level has already risen by about 25 centimetres in the last 100 years, and while most of our beaches are doing okay, there are many erosion 'hotspots' where houses and roads are literally falling onto the beach. Hindsight is always 20/20 and it's hard to argue with decisions to live close to the coast in the past, but it's important that we learn from these mistakes for the future sustainability of our beaches by making sensible planning decisions when it comes to the coast. One sensible decision is to make sure that 'setback' zones, which are designated areas behind shorelines where building development is not allowed, are based on some of the higher sea level rise predictions. When it comes to beaches and human development, it's always best to err on the side of caution.

## BEACHCOMBING AND BEACH FOSSICKING | The ocean works in mysterious ways.

I once left my favourite thongs on a remote Australian beach one night and went looking for them the next morning only to find that the tide had come up, leaving one, but cruelly taking away the other. A few months later I was beachcombing in Fiji and found a thong almost identical to my missing one. This gave me a working pair again because, for sentimental reasons best left unexplained, I had kept the surviving thong. The point is, waves and tides bring in all sorts of things to the beach and you never know what treasures you might find each day.

Beachcombing isn't hard. It involves walking along a beach and looking at what's been washed up. It could be shells, a piece of driftwood that makes the perfect ornament, exotic glass fishing floats from far away oceans, or a shipwreck exposed by storm waves. Or you can spend your time at the beach looking for neat pebbles and stones. This is called beach fossicking. You might find greenstone (a type of jade) washed down from mountain rivers on west coast beaches of New Zealand's South Island, scour the shores of Lake Superior in the Great Lakes of North America looking for semi-precious agates (better yet, try Agate Beach in Oregon), or make some jewellery out of sea glass – broken bits of glass smoothed and frosted by the tumbling action of waves and sand.

Be careful though, some of those rocks look amazing when wet, but pretty plain when dry. Beach fossicking can also be expensive. During my backpacking days, when most of my friends were sending home exotic sarongs and artwork, I sent home boxes of beach rocks. The postage was astronomical. My all-time favourite is a big blue rock from the Indonesian island of Flores. It's still blue and I have no idea what it is, but it looks great and brings back memories and that's all that matters to me.

# The bottom line

- ★ It's all about the rocks. Geology is the most important control on what your beach is made of and where it is.

- ★ Sand is made of minerals brought to the beach mostly by rivers.

- ★ The colour of beach sand depends on the type of mineral in the rocks; quartz tends to be lighter in colour, heavy minerals are darker.

- ★ Beaches are really made up of a wide range of different types of sediments that have different sizes and shapes.

- ★ Common types of sediments found on beaches include sand, gravel, mud and carbonates.

- ★ The smaller the size of the sediments on a beach, the older it is and the further the sediment has travelled to get there.

- ★ Sea level has a huge impact on the amount of sand and sediments available to make beaches.

- ★ Sea level rise results in a shortage of sand and moves beaches inland.

- ★ Most of the world's beaches have been in their present location for about 6500 years.

- ★ Black sand can get really hot on sunny days!

# 2

# Waves

## The science of the surf

When you get a chance, search online for 'biggest wave ever ridden' and you'll no doubt end up watching some jaw-dropping footage of surfers the size of ants riding some incredibly monstrous waves (Pic. 6). While most of us will never know what that feels like, anyone who's ever surfed, boogie boarded or bodysurfed a largish wave and miscalculated their take-off will be familiar with the sensation of time slowing down as you are briefly suspended in mid-air. It's amazing what you can accomplish in the millisecond before you plunge 'over the falls'.

You can easily estimate the distance you are going to fall, briefly consider how many tonnes of water will soon be impacting on your head, instinctively distance yourself from your board that has now transformed into a waxed impaling device, and still be able to sum up your predicament in a simple two-word statement, 'Oh-oh'. Or something to that effect. But how many people would have the presence of mind to wonder how far the wave travelled to get there or the physics of the wave-breaking processes that got you into this mess? Surprisingly, if they are experienced surfers, probably quite a few.

Understanding how waves at the beach work isn't always a life-and-death matter, but it can definitely help. You don't have to be a surfer to be interested in waves either. Lifeguards, swimmers, parents, fishermen, scuba divers and people walking along the beach, anyone who loves the beach really, should all know a little bit about what's going on in the water. The beauty of waves is that the science of the surf can be quite simple and will help you choose the right wave and avoid the wrong wave the next time you step into the ocean.

Or you can just appreciate waves on beaches as the beautiful and powerful forces of nature that they are. This chapter deals with some of the basic questions about these amazing motions of the ocean, particularly how waves form, where they come from, what makes them break, what are safe and dangerous waves and, most importantly, how we can have fun on them.

# Wave formation

## The answer is blowing in the wind

Throw a pebble into a pond and instantly lines of small waves will start radiating away from the splash. Waves in the ocean aren't normally created like this, but the point is that something needs to disturb the water surface in the first place in order for waves to form. Imagine a harbour or bay on a completely still day. The water is flat calm. Now imagine the same harbour during a gale. The water is awash with messy, choppy, whitecapping waves (Pic. 7). It shouldn't come as a great surprise to learn that wind is what disturbs the water surface and creates waves. It doesn't matter if it's a puddle or the Pacific Ocean, when wind blows, waves will form, but the growth of ocean waves from tiny little ripples to huge 10-metre swell is much more interesting than that.

If you have a fish aquarium at home, lean over the tank and blow across the water surface as parallel to it as possible. Not much will happen. Now try blowing downwards. The force of your air will physically push the water downwards creating a lump that will start moving to the other side of the tank. Congratulations, you have just created a wave by transferring some of the energy from your breath into the water! This is an important concept – the reason waves move is because they are carrying energy.

The same thing happens in the ocean, but it's easier to create waves here because the ocean surface is rarely completely flat and smooth. There are always little bumps and undulations around for all sorts of reasons. This means that when wind gusts blow across and down on the surface of the ocean, they have more to grab onto, which helps transfer energy from the wind to the water.

As wind blows across the ocean (or any body of water), the first waves that form are really just tiny ripples, but they make the ocean surface even rougher and the wind can transfer even more energy into the water. As long as the wind keeps blowing, the little waves will start to grow in size, move faster and start to merge together with some of their older relatives. In this way, they keep getting bigger and bigger. If the wind only blows for a few minutes though, the energy supply is shut off and the waves will peter out and disappear.

However, if the wind keeps blowing for hours and even days and maybe gets a little stronger, the waves will continue to grow and a snowball effect starts to kick in. As the ocean gets rougher, the wind has more and more bumps to grab onto and pump even more energy into the ocean, so the waves grow and grow and grow. There's almost no stopping them. At the same time, the classic wave shape begins to develop. The high points of waves are called crests and the low points are troughs. The vertical distance between the crest and a trough is the height of the wave. In strong winds, flat calm water conditions can turn to choppy waves very quickly. If you've ever been on a kayak or small boat struggling to stay afloat in a sudden wind squall, you will know just how fast waves can grow!

## What controls wave height?

It may be tempting to conclude that as the wind keeps blowing, the waves will keep getting bigger and bigger until every holiday cruise turns into *The Poseidon Adventure* and tow-in surfing becomes a daily event at our beaches. Fortunately this doesn't happen because a number of factors limit how big waves can grow, as shown on page 32. The first is the wind speed. The faster wind blows, the bigger waves will be. The second is the wind duration, or how long the wind blows for. If it's just a few seconds or minutes, only tiny waves are formed. If it blows for hours and days, the waves will be much larger. Third is the fetch. 'Fetch' is a sailing term used to describe the distance over water that wind can blow without being interrupted.

A small lake may have a fetch of only a few hundred metres and even if hurricane-force winds blow over it for months, it still won't develop big waves because waves need

# THE BIGGEST WAVE EVER SURFED

Big wave surfing and Hawai'i are virtually synonymous. Not only was modern surfing pioneered in Waikiki by Duke Kahanamoku in the early 1900s, big wave surfing also began on Oahu's North Shore at places like Waimea Bay. So just how big is big? In 1969, Australian Greg Noll caught an 11-metre wave at nearby Makaha Beach that was considered the largest wave ever ridden. He promptly retired from surfing! This 'record' was consistently broken once surfers started getting towed out by jet skis onto giant waves at a break called 'Jaws' in Maui and surfing 20-metre waves, but it's always been contentious determining the biggest wave ever surfed because most wave heights were based on visual estimates.

Since then, a number of dedicated (and possibly crazy) big wave surfers have sought to conquer extreme surfing spots like Teahupo'o in Tahiti, Shipstern Bluff in Tasmania, Mavericks and Cortes Bank in California and the newest big wave mecca: Nazaré, Portugal (Pic. 6). Do yourself a favour and watch some footage from Nazaré. It's frightening! The record for the biggest wave ever surfed is now a 26.2-metre monster ridden at Nazaré by German surfer Sebastian Steudtner (where is the surf in Germany you may ask?). Will this record be broken? Of course! Big wave surfers have their own holy grail – the 100-foot (30.5-metre) wave and who knows, maybe it's already been surfed.

# HOW TO MEASURE WAVE HEIGHT

**W**ave height is the vertical distance between the crest and trough of a wave. While most surfers are pretty good at estimating wave height consistently and accurately, there also seems to be a lot of underestimation going on. I've watched fully grown adult surfers riding walls of water towering over their heads while other surfers standing next to me are calling the waves '4 foot'. A non-surfing observer might conclude that most surfers are on the short side, need glasses and have completely rejected the metric system. In truth, while there may be a bit of machismo and one-upmanship involved in casually underestimating wave height, the real problem are the Hawaiians. Apparently the early Hawaiian surfers gauged wave height from the back of the wave, so that they were only really seeing half the true wave height. Over time, this method has spread around the world.

How we actually measure wave height is a different story. For years, all we had were shipboard observations made in the middle of the ocean. More recently, wave rider buoys, which look like big beachballs with pointy antennas sticking out, have been tethered to the seabed several kilometres offshore and record wave height continuously as they pitch, bob and roll about as the

Crest

= Wave height

Trough

waves pass them by. However, there are not many of them around and they only cover a small area. Nowadays we look towards space at satellites that estimate wave height by firing radar beams down at the ocean and measuring their reflection off the sea surface. Sounds promising, but if you really want to know the wave height before you jump in, use some old technology and ask a surfer. They're more accurate, a lot cheaper and still use the imperial measuring system.

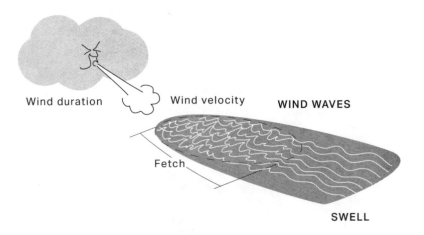

Wind duration     Wind velocity     **WIND WAVES**

Fetch

**SWELL**

The formation of waves by wind is controlled by the
wind speed, the duration of the wind and the distance
the wind blows over water. Over long distances short
and choppy wind waves eventually turn into swell waves.

room (a long fetch) to grow. In contrast, the middle of the
Pacific Ocean has a fetch of thousands of kilometres, so the
wind and waves have plenty of room to blow and grow. So
the size of waves at your favourite beach not only depends on
the weather and climate, which produce the wind, but on the
geology and shape of the continents and ocean, which control
the fetch.

## Motions in the ocean
### Wind, sea and swell

On surf beaches around the world, there is an unofficial
society called 'The Dawn Patrol'. Surfers hit the water early
and experience more sunrises than insomniacs, but there's a

reason for their dedication. Mostly it's because they have to go to work, but mornings also tend to be less windy and the waves are smoother and glassy and better for surfing. There seems to be a contradiction here though. There's no wind, but there are still waves? So where did the waves come from? The truth is that all waves are formed by wind, but there are different types based on how far they've travelled.

Anyone who goes to the beach is familiar with the refreshing onshore sea breeze that kicks in on hot summer days around midday or early afternoon. Sea breezes occur because on sunny days, the land heats up faster than the ocean so air over the land will start to rise and cooler air blows in from the ocean to replace it. While the sea breeze provides welcome relief to beachgoers and coastal dwellers on scorching hot days, surfers hate it because within minutes the surf is turned into a choppy mess of whitecapping waves that are almost unrideable and it can also make swimming conditions unpleasant. As these waves are generated locally near the beach they are called wind waves because you can see them being formed by the wind. They are also common in protected and smaller bodies of water, such as harbours, bays and lakes, whenever strong winds occur (Pic. 7).

Once formed, all waves travel at a certain speed and that speed is related to the wave length, which is the distance between two wave crests. Working out wave speed is quite easy thanks to basic wave physics and theory, which I will conveniently ignore here. When you are driving your car, your speed is in kilometres per hour. So speed is just a distance over time. In terms of waves, wave speed can be worked out by dividing their wave length (a distance) by the wave period (a time). The wave period is the time it takes for two wave crests to pass the same point. Both wave length and wave period can be measured pretty easily so it's simple to work out how fast waves are travelling. That's why wave prediction is getting better and many surfers check swell forecasts on the internet every day.

Now imagine a chaotic mess of wind-generated sea waves in the middle of the ocean. Within the mess there's a whole bunch of waves that have different heights, periods and wave lengths. Because all waves move, they will all start to travel away from the wind, or storm, that created them in all different directions, but what happens is that the longer waves will always travel faster.

So as the waves are travelling, not only are they smoothing and separating themselves out, but if they have a huge fetch of thousands of kilometres to travel across, they will start to turn into swell waves with longer and straighter crests that are much smoother and cleaner in shape (Pic. 8). They also have periods anywhere from 8 to 20 seconds and the further the wave has travelled, the longer the wave period will be. Very long swell waves are called groundswell and can literally appear out of nowhere on beautiful days without a trace of wind. That's because they were formed way out at sea by a storm far away and have finally arrived at a beach after a journey of several days or more.

Swell is purely an ocean phenomenon. Smaller bodies of water just don't have enough fetch to generate swell, not even the Mediterranean Sea or the Great Lakes of North America. But while some smaller bodies of water will only experience wind or sea waves, large oceans can have a mixed swell and sea wave environment. Many exposed coastlines will always have some swell approaching from one direction or another, but if there is a localised intense weather system around, such as a cyclonic depression, choppy sea waves will form on top of that swell. This is why marine forecasts can report a 1-metre sea on a 2-metre swell.

How far waves travel also depends on how much energy they have. Most wind waves don't last very long because once the wind stops, the waves don't have enough energy to keep travelling on their own and they die out. Sea waves are a little different because there is an exponential relationship between the height of a wave and how much energy it has. Little increases in wave height result in much bigger increases in wave energy. Sea waves tend to be a little bigger, so have more energy and can travel further once the wind has stopped or they have moved away from the storm. Once waves turn into swell, they generally have so much energy that they can travel fully independent of any wind. Swell also doesn't lose much of its energy as it travels and it is not uncommon for individual waves to travel over 10 000 kilometres before breaking on a distant shoreline!

While it's easy to understand that waves carry energy, you often hear surfers talk about some waves being more powerful than others, particularly in places like Hawai'i. You might just brush it off as some sort of surfer lingo, but wave power is real. I remember bodysurfing at Slaughterhouse Beach in Maui, Hawai'i and the waves looked pretty much like they did on a normal day back in Australia. No big deal. The first wave I caught felt like someone had injected me with rocket fuel! I just took off and it was crazy fast and I could almost feel an extra sort of force pushing me. I think my eyeballs were like saucers. I suddenly understood what surfers meant by wave power.

So what's going on in Hawai'i? Why are the waves so powerful. A technical definition of power is that it's the rate of doing work, or delivering energy. Wave power is therefore the product of the energy of the wave and the speed at which the wave is travelling. That makes sense. It basically says that the higher the wave and the faster it's moving, the more powerful it will be. But what controls wave speed? The main control on wave speed is the wave period, meaning that longer period waves (with longer wavelengths) travel faster. Hawai'i sits in the middle of the Pacific Ocean and gets long period swell from every direction due to the long fetch. That long period swell is the key to wave power. I even remember boogie boarding the famous Banzai Pipeline on the North Shore of Oahu – before you think that's impressive, it was only about 30 centimetres – and still feeling that power. Amazing. It's not just a phenomenon in Hawai'i though. The same thing happens on any beach when a long period swell, often called a ground swell, with periods of 12 seconds or more rolls in.

## Wave sets: every seventh wave is *not* a big wave

If you spend some time at the beach looking at the waves, it will become pretty obvious that the waves rolling in are not all the same size. Some are bigger, some are smaller and sometimes there are flat spells. My parents told me that every seventh wave was a big wave. I used to test out this theory by counting the waves. It never worked. I would lie on my inflatable raft counting and counting, waiting for the big one and nothing would really change until, all of a sudden, about five huge waves would come out of nowhere and wash me up on the beach. So no, the idea that every seventh wave is a big wave is not true, but like many old sayings, there is an element of truth to it and that element is something called a *wave set*.

A wave set is a group of several large waves that seemingly appears out of nowhere. A single group can consist of anywhere from three to ten larger waves. Not only does the number of waves in each set vary, so too does the length of time between sets and the height of each wave in the set. So what's going on? Well, it's worth remembering that the oceans are pretty big. There can be a number of storms occurring at different places at different times creating multiple waves moving across the ocean with all sorts of different wave heights, wavelengths, periods, speeds and directions. In other words, there's a lot of wave traffic out there.

When different waves come together a whole lot of addition and subtraction starts happening. If the crests coincide, the new crest is bigger. If the troughs coincide the new trough is deeper. This is called *constructive interference* because the crests and troughs have added together. However, if the crests and troughs of the wave trains overlap, the whole thing gets cancelled out and the result is a flat spot in the resulting wave. This is called *destructive interference*. If you start to add more waves into the mix the resulting wave patterns get even more complex. Once the waves get locked in with each other though, they travel as a new wave that will have some sections of constructive interference (the large waves of the wave set) and some sections of destructive interference (the lulls between sets). This simple explanation illustrates why sets are so varied and seemingly random in so many ways and why surfers sometimes have to wait ages to catch the best waves of the day. They're a patient bunch.

USING THE BEACH

## THE ART OF BODYSURFING |

Bodysurfing is the purest form of ocean wave riding because it's just you and the wave and is a fantastic way to stay fit. It should also be easy, but often you'll see people catching waves effortlessly while others flail and flounder away. Bodysurfing is all about location, timing and wave selection. Before you head out, watch what the surf is doing. Obviously you need to go where waves are breaking, but avoid areas that are too deep where the waves are barely breaking and give the shallowest areas a miss because you don't want to be slammed into the sandy bottom.

Look for a nice clean face of a steepening wave just before it breaks. If it's not steep enough, you won't catch it, or it won't break. If it's already broken, you'll just get swamped with whitewater. When you see the right wave approaching, swim hard towards the beach so that your own speed will closely match that of the wave just before it begins to break. Don't stop kicking your legs and paddling with your arms until you feel yourself sliding down the front of the wave. You can then ride it straight or try to angle across it, going left or right depending on which direction the wave is peeling. Always keep one or both arms held out in front of you.

Purists will argue that using flippers and hand paddles is cheating, but they definitely help. Flippers (or fins) allow you to build up your launch speed more quickly, helping you catch more waves. They also help you swim back out more easily. The paddle provides lift and with your paddle arm stretched out, you can steer along the wave and perform all sorts of different manoeuvres. At the same time you can use your free arm to windmill to keep up your speed and also provide more stability on the ride. Bodysurfing takes practice so expect to miss a few

Bodysurfing with flippers at Cape Solander, Sydney, New South Wales.
*Wikimedia Commons user Ian Macdonald*

waves and get wiped out by others before you are successful consistently.

Your arms serve an important role in bodysurfing. Not only do they help support your body and steer while riding the wave, they also take the impact if you happen to hit something. Unfortunately, many people bodysurf with their arms held behind their body with their heads sticking out like a Pez dispenser. It's a good way to smash your face and neck into other swimmers, surfboards and the hard bottom. It doesn't matter if you've been doing it that way all your life, it only takes one mishap to send you to the spinal ward. It's better if your arms hit the bottom rather than your head. If you don't believe me, ask the pros. They do it.

# Wave motion: don't fall asleep in the deep

Waves go up and down. If you've ever spent time lolling away the day floating on an inflatable raft in the ocean, you'll be going up and down as well and probably still will be when you hit the pillow that night. I'm not sure why this happens, but my totally unscientific and unproven theory is that the water in your brain is still sloshing around. If you've ever scuba dived in wavy conditions, you will also know that waves move you backwards and forwards, but this motion lessens the deeper you go until it eventually disappears and you don't feel the waves at all.

Sometimes this to-and-fro motion can be quite pleasant. So pleasant in fact that while conducting a beach experiment in Georgian Bay, Ontario, Canada, that involved a lot of scuba diving and evening recreational activity, I was so tired on a morning dive that I actually fell asleep for 15 minutes! Given that most tanks only hold about 45 minutes of air, I wouldn't recommend taking a nap

*Waves exist to carry energy and it is the energy that moves and creates the wavy shape, not the water.*

while diving, although I did feel better afterwards. For others, the to-and-fro motion is not such a pleasant experience. Wave motion can lead to seasickness, both on the surface and even underwater. Although the latter hasn't happened to me, it's unpleasant to watch. Extremely unpleasant.

We all know what waves look like and that they seem to be moving. But what is actually moving? Waves exist to carry energy and it is the energy that moves and creates the wavy shape, not the water. It's the same for light waves, sound waves and even Mexican waves made by spectators at sporting events. The people may jump up and down and throw beer cups to form the wave, but they don't actually move anywhere. Meanwhile the energy of the wave goes around and around the stadium. The ocean is no different. The wave energy travels through, but the water particles move in circular orbital motions underneath the wave. Imagine floating on a raft again. You will drift slightly onshore with the crest of the wave, downwards as the crest passes, back offshore in the wave trough and then up again as the next crest approaches. In a sense you've completed a circle, but you really haven't gone anywhere. So the motion of the wave energy and the motion of water in a wave are two different things.

You feel these orbital motions more strongly near the surface. In deep water, they get smaller in diameter the deeper you go, to the point where they disappear completely. However, when waves travel from deep to shallow water, at some point the orbital motion of water particles underneath them starts to hit the seabed, causing the wave to slow down because of friction, a process known as 'feeling the bottom'. Despite eliciting giggles and snorts from my university students, the fact that waves 'feel bottoms' is extremely important because that's what causes them to change shape as they get closer to the beach.

If you look out to sea from a headland, the waves offshore are barely noticeable, but as they approach the beach they seem to bunch together, their crests become sharper and their troughs stretch and flatten out. This whole process of waves slowing down, becoming steeper and increasing in height is referred to as wave shoaling and happens to every wave that enters shallow water. Just how waves shoal eventually deter-mines how and where waves break and whether there'll be dec-ent waves or not and it all depends on the shape of the seabed.

## BOOGIE BOARDING
## FOR YOUNG AND OLD

**WHAT IS A BOOGIE BOARD? |** Cut the front half of a surfboard off, make it flatter, thicker and wider, replace the pointy nose with a broad snout, replace the fibreglass with styrofoam and you've got a boogie board. The idea is that you can lie on it with the front half of your body supported and hold onto the front with your arms, while your lower half dangles over the end allowing your legs to kick freely. This allows you to build up speed to catch more waves and ride longer distances than you can bodysurfing. It's very easy to learn and you can become proficient after just a few sessions, but once you start taking it seriously, it's no longer called a boogie board, it's a 'bodyboard'!

**WHAT DO YOU NEED? |** Boards range in price from dirt cheap to hundreds of dollars. You get what you pay for. There are two other accessories that are crucial. One is a wrist strap and leash, which keeps you and the board connected if you get knocked off. The other is a pair of boogie boarding flippers because: BOOGIE BOARDER + NO FLIPPERS = IMPENDING RESCUE. Not only will you struggle to catch waves without them, if you end up in a rip current, you might as well put up a sail and head offshore. It's so much better going boogie boarding with flippers.

**PADDLING OUT |** To avoid flapping around and falling flat on your face while walking with flippers on, try putting them on in the water or walk in backwards wearing them. Staying on the board without sliding off can be tricky at first. Wearing a wetsuit or rash vest gives you a bit more grip and don't lie too far back on the board. Paddling with your arms and kicking with your legs helps build up a good rhythm and speed. Arch your back and look straight ahead with your head up. If a broken wave is rushing towards you, bend the nose of the board down under the surface and push under the wave. This is called duck diving. After the wave has passed, you and the board should pop back up and you can keep paddling.

**CATCHING A WAVE |** Wave selection and timing is the same as for bodysurfing, but it's not just a matter of choosing a wave and pointing the board towards shore. You need to kick hard with your fins to build up speed. Paddling with one arm also helps, but keep the other on the board for stability. Once you catch the wave, you can ride it all the way to the beach, but that means a long and hard paddle back out. If the waves are peeling left or right, you can ride sideways along the wave and might even tuck into a 'barrel', otherwise known as the hollow part of the wave. Avoid waves that are plunging and closing out. Once you master riding the waves, you can take it to the next level and start doing 360s, somersaults and riding on one knee. Or you can buy a surfboard.

# Wave breaking

## What lies beneath is really important

If the bottom of the ocean was completely smooth and sloped gently upwards towards a beach that was perfectly straight, the surfing would be awful. All the waves would approach the beach in a straight line and break at the same time as a 'close-out'. Throw a few sandbars, channels and rock reefs into the mix though and, all of a sudden, things get interesting because the wave crests will start to bend. This bending is called *wave refraction* and is what makes a good surfing break (Pic. 9).

Wave refraction occurs whenever one part of an incoming wave crest hits shallow water first. That part of the wave will slow down while the rest of the wave travelling in deep water still moves quickly. The different speeds along the crest are what cause the wave to bend. Surfers can then catch a wave just starting to break and ride the bending wave all the way to the beach. This is basically what a point break is. Most point breaks are prominent rock reefs sticking out into the ocean, which cause waves to peel around them. If the wave is breaking towards the right, it's called a right hander; if it breaks towards the left, it's a left hander. If you are into long rides, try visiting Chicama in Peru where the wave can break for 4 kilometres and paddling back out is not an option! Instead it's an awfully long walk back along the beach to start again. Pavones in Costa Rica and Manu Bay in Raglan, New Zealand, also claim to have breaks of up to 2 kilometres long under the right conditions.

Surf breaks caused by wave refraction can be quite predictable where the bathymetry (the depth of water) is fixed by local geology. Surf breaks that depend on features such as sandbars, which can shift around on a daily basis, can be more fickle. Unfortunately, waves are far from consistent and there can be months of dreary surf, depending on the weather and the shape of the seabed. Also complicating matters is that waves can approach a coastline from all sorts of different angles and each angle can produce different refraction patterns. For example, a surf break that may produce perfect barrels on a northeast swell may be almost unrideable on a southeast swell. The famous big wave surfing spots around the world that produce frightening images of tiny surfers dwarfed by mountains of water (Pic. 6) only work because their underwater bathymetry is tuned to certain extreme patterns of swell conditions. In other words, what's underneath is really important.

## Why do waves break?

Depending on your swimming ability and experience level, dealing with waves as they break can be a fun or challenging experience. Whether you are catching a wave, or are scared out of your mind at the looming wall of water bearing down on you, knowing why and how waves break is very useful information. Waves break when they get too steep. It's that simple. As they enter shallow water and start to slow down, they get steeper and steeper and steeper until something's got to give. And it does. The wave breaks. A decent ballpark estimate is that waves heading towards the beach break when the water depth is about 1.3 times the wave height. In other words, a 1-metre-high wave will break in about 1.3 metres water depth and a 3-metre wave will break in about 4 metres of water. What this relationship also means is that bigger waves will break further offshore.

# Types of breaking waves

Just as the shape of the seabed can vary enormously, so does the way in which waves break and not only do breaking waves look completely different, some are more dangerous than others. Here are the three most common types.

## Plunging waves

For those old enough to remember, think back to the intro to the TV show *Hawaii Five-0*. Otherwise just imagine a beautiful image of a wave curling over forming an almost perfect hollow tube. That's a plunging wave. Also known as dumpers, barrels and 'the green room', these are probably the most famous breaking waves because they are the most dramatic and beautiful (Pic. 10). They are also the most dangerous because they curl over and crash downwards with an explosive force. Almost everyone who's gone to a beach with waves breaking at the shoreline has been 'blown up' trying to get in the water at some point!

Plunging waves occur whenever waves travel from deep water to shallow water over very short distances. This makes them slow down so fast that the curl you see is really the orbital motion of the water particles overtaking the speed of the wave. They are very common on sandbars, rock reefs and at the shorebreak of steep beaches and are exciting to surf, but it's important to remember why they 'barrel'. There's something hard, shallow and unforgiving underneath. If a plunging wave is about to break on you and you are trapped in no man's land, getting under the wave is the best thing you can do. Dive under the water towards the wave with your arms held out in front and stretch your body flat against the bottom and hold on to the sand.

### Spilling waves

On fairly flat beaches, waves will slow down gently, steepen gently and break gently as the crest spills, or crumbles, down the front face of the wave. Spilling waves are often called rollers because the whitewater tends to cascade along as the wave keeps travelling (Pic. 11). Although they can be turbulent, if you see a spilling wave coming your way, you don't need to panic. All you have to do is duck your head under the water surface and 'whoosh', the whitewater will rush over you. Most times you'll barely even feel it.

Without a doubt these are the safest waves to surf on. Don't be fooled though, they can still get big. If a huge spilling wave is tumbling your way, do the same thing as you would for a plunging wave: dive under and hold onto the bottom. If you open your eyes while doing this, you can also see some amazing wave patterns and reflections, albeit a little blurry.

### Surging waves

Some waves are very sneaky. They don't plunge or spill, but just bulge up near the shoreline and then collapse by suddenly rushing up the beach (Pic. 12). Surging waves are not generally large in height, but can pack a surprising wallop, knocking people over or causing them to lose their footing. They tend to occur more often on beaches that are moderately steep. They are also associated with strong swash action. Swash is a term used to describe the uprush of water up the beach when

# WHAT MAKES A
# DANGEROUS WAVE?

It depends who you talk to. Only adrenalin junkies would really enjoy surfing waves over 5 metres high and even they'd admit there's danger involved, but waves don't have to be big to be scary. A perfect 1.5-metre wave to an adult surfer takes on an entirely new meaning to a six-year-old staring up at it. Plunging waves are the most dangerous and a common source of neck and spinal injuries because they have a tendency to slam dunk people onto the shallow and hard bottom underneath. It's an explosion of energy and what you don't want is that energy exploding on your head (Pic. 13).

Surging waves are generally not that big, but they can rush up the beach with great speed knocking both children and adults over like bowling pins and then rolling them back down the beach like bowling balls. As the water rushes back down the beach, it can pull people under the water and offshore a little way as well. If you've ever heard of the term 'undertow' before, this is probably what people are referring to, but the good news is that because we all have a natural tendency to float, you never stay under for more than a few seconds.

Ultimately, whether a wave is dangerous or not really depends on your ability to recognise hazardous wave conditions and to know not only your limits in the surf, but those of family and friends around you. Any wave is potentially dangerous.

waves break followed by the backwash of water returning back down the beach. All waves turn into swash when they reach the shoreline. Just how strong the swash is depends on the type of breaking wave and the slope of the beach and it can be unpredictable so always keep an eye on your kids!

If you've spent a fair bit of time at the beach, swimming or just watching the waves roll in, chances are you've seen all of these different breaking wave types. One thing not to forget is that wave breaking is related to water depth, which means that on ocean beaches with tides, the location and type of breaking waves will also change with the tide. How waves break is obviously important for surfers, but it is also important to parents supervising children at the beach who need to know just how safe the surf conditions are and how they and their kids should behave in the water.

## There is no such thing as a freak wave

Tell that to the odd cruise ship and supertanker that gets flipped around by a monster wave that seemingly pops out of nowhere. Freak, or rogue, waves definitely occur in the middle of the ocean. It's all about the law of probability. Every now and then, the right combination of waves will meet from different directions and combine together in one brief moment of random constructive interference to form a giant wave that appears out of nowhere and disappears just as quickly. They probably occur more commonly than we think, but there's no one around to see them.

Closer to the beach, there are often reports of people being swept off of nearby rocks and rock platforms by freak waves, even when the waves are fairly small. In this case the 'freak' waves are likely wave sets or reflected waves. When waves hit a rock, reef or headland, they can literally bounce back offshore. Sometimes the reflected wave combines with an incoming wave to form a sudden giant wave, which can be extremely dangerous to anyone on the rocks (Pic. 14). It's very much about being in the wrong place at the wrong time. Once someone gets knocked in the water under these conditions, it's often hard to get back out because of all the reflection and turbulence. It's no surprise that in terms of fatalities, ocean rock fishing is considered to be one of the most dangerous sports in the world.

Do freak waves happen on beaches? Sort of. Have you ever been lying on your towel on a beach, seemingly high and dry, when suddenly a surge of water rushes up the beach and swamps you? This is something scientists call surf beat, but it's commonly referred to as sneaker waves. Sneaker waves can be surprisingly freaky and strong enough to get people into trouble by knocking them over and dragging them back down the beach. They tend to form on high energy beaches that consistently have large wave conditions, or during storms. All that wave breaking releases a lot of water and energy that starts sloshing around in the surf zone and building up so that every now and then it forces a big rush of water up the beach like a surging wave on steroids. Many beaches on the west coast of the United States post warning signs about sneaker waves and the notoriously dangerous Reynisfjara Beach in Iceland is extremely popular with tourists, but despite the presence of a flashing light warning system, sneaker waves have claimed multiple lives in recent years. I was lucky enough to visit in very calm conditions and even then, sneaker waves had people scampering up the beach to safety.

My friend Phil Osborne was doing a beach experiment at Muriwai Beach in New Zealand and had his computers

LEARNING TO SURF | Let's face it, learning to surf is a lot easier when you are a young kid or teenager when your body is fit, loose and flexible and you don't care about things like falling off and having the board conk you in the head. Having said that, it's never too late to learn. Persistence is the key and before you know it you'll be up and riding. At least that's what I keep telling myself. Bodysurfing just seems so much easier! Boardriding is a whole different kettle of fish so here's some tips from my surfing buddies.

**GET FIT |** Surfers are extremely fit as paddling into waves and back out again can be very strenuous. Make sure you're in good shape before you start, otherwise you'll give up quickly because you'll be too exhausted. Kids don't seem to have this problem.

**USE THE RIGHT BOARD |** Start with a longer, wider board because it provides a more stable platform to stand up and work out things like balance and turning. Don't be fooled into buying a cool-looking short board covered with nifty designs and stickers. After endless frustration, you'll end up selling it for much less than what you bought it for. Soft boards, or 'foamies', not made out of fibreglass are worth considering because they hurt less and are cheaper. Oh, and don't forget that leg rope.

**WETSUITS AND WAX |** You need to put a lot of practice in when learning and unless you're in the tropics, you'll quickly get cold. Get a wetsuit that fits snugly and is nice and flexible, allowing you to paddle freely with your arms. A steamer is a full-length suit that comes in different thicknesses and is good for water under 18°C. Once the water gets to about 14°C you'll need at least a 4-millimetre-thick suit. A spring suit has short arms

USING THE BEACH

and legs and is good for summer and warmer water. Rash vests are a good idea to wear underneath to reduce chafing from the wetsuit. Getting savvy with boardwax is advisable as you will need all the grip you can get.

**FIND THE RIGHT WAVES |** There is no point starting off in a large swell that is producing perfect barrels. You will not catch these waves for a long time and will only end up going headfirst over the falls and having your board snapped like a twig. Swallow your pride and head to a beach that is nice and flat and start catching the whitewater from small, gentle, spilling waves. Whitewater from broken spilling waves is easier to catch and once you stand up, you can ride them quite a long way, allowing you to work on your balance and turns. Once you get a bit more confident, you can try catching 'green' (unbroken) waves.

**PADDLING AND STANDING UP |** Paddling is pretty straightforward. You need to paddle hard with your arms to build up speed and it's similar to swimming. Make sure you are not too far forward on the board so that you are pushing the front of the board down, and not so far back that the front is sticking up. As you feel the board angling down and starting to catch the wave, grip the sides of the board around chest level and push yourself up while springing to your feet with your best foot forward in one fluid motion. Practice makes perfect!

**DUCK DIVING |** When paddling out you will face waves coming at you with surprising ferocity. If you just lie there, they will smash into you, pushing you and your board all the way back to the beach or even worse, pushing your board into your face. This can be frustrating, so you need to take evasive action. Just before the wave hits, grab the sides of the board with your hands, keep your arms straight and push the front of the board down into the water at a sharp angle. At the same time, bring up one of your knees to the board and in one movement push yourself under the incoming wave and then flatten out on the board once you are underwater. After the wave passes, angle the front of the board upwards and your buoyancy will bring you back to the surface in a perfect position to keep paddling.

**KNOW THE RULES** | There are rules in the surf and if you don't follow some of them you might get to experience some surf rage. As a learner, it's best to stay away from crowded surf breaks full of experienced surfers and locals. Once you start riding waves consistently, the key thing is not to drop in on somebody. In other words, don't cut them off. Before catching a wave, have a glance around. If a surfer is already on the same wave riding it in your direction, it means they have right of way. Pull back and wait for the next turn. There are plenty of waves to catch.

**KNOW THE LINGO** | If you pick up a surf magazine for the first time, it's like reading a foreign language. There's 'lefts' and 'rights' and 'barrels' with people 'carving' and 'peeling' and 'goofy footing' and sometimes the surf is just plain 'sick'. It's no good asking anyone in the 'line-up', they'll just ignore you. Try googling to improve your 'surfese'.

**DON'T GIVE UP** | Learning to surf is not going to happen overnight. You might stand up a few times during your very first attempt on small spilling waves, but to become proficient takes time and practice. Put in the hard yards though and you will eventually become a surfer.

and bunk beds in a camper van halfway up the sand dunes. The waves were consistently breaking about 100 metres away down the beach. One night they were completely swamped by a single wave and everything was trashed. When they went outside to have a look, the waves were still breaking further down the beach. Phil now believes in freak waves. Then again, maybe it was a tsunami!

# The bottom line

- ⭐ Waves are caused by wind.

- ⭐ The stronger the wind blows, the longer it blows for; and the longer the distance over water it can blow, the bigger the waves will be.

- ⭐ Wave height is the distance between the crest and trough of a wave, and wave period is the time for two consecutive crests to pass the same point.

- ⭐ Wind waves are formed by local winds near the shoreline and are short, choppy and messy with periods of about 3 to 8 seconds.

- ⭐ Sea waves are bigger wind waves formed by stronger winds in the middle of the ocean.

- ⭐ Swell waves start off as sea waves, but have travelled very long distances and hit the beach as long crested, smooth waves with periods of 8 to 20 seconds.

- ⭐ Wave sets are groups of larger waves that occur randomly in most swell wave conditions.

- ⭐ The shape of the seabed and the beach causes waves to slow down, change shape and break as they go from deep to shallow water.

- ⭐ Waves can break mostly by plunging, spilling and surging, with plunging being the most dangerous wave and spilling being the safest.

# 3

# Motions of the oceans

## Tides, tsunamis and storms

On 26 December 2004 an earthquake registering 9.3 on the Richter scale occurred off the west coast of the Indonesian island of Sumatra. During the earthquake, a 1200-kilometre section of seabed ruptured and was pushed approximately 20 metres upwards into the ocean above. This was the disturbing force that displaced a massive slab of water, creating very long and fast tsunami waves that travelled at speeds of up to 500 kilometres per hour across the Indian Ocean in all directions. The waves smashed into beaches and coastlines in their path, reaching estimated heights of 30 metres above sea level in some places in Sumatra. The tsunami caused an estimated 300 000 deaths, untold billions of dollars of damage and, in some cases, completely altered the shape of the coastline.

Although it was the closer Indian Ocean beaches that were dramatically affected, traces of the tsunami waves were detected on tide gauges in every ocean around the world. Catastrophic tsunamis clearly have far-reaching effects and can impact almost every ocean beach. They are impossible to predict and, not only have destructive waves of this magnitude happened before, they will happen again – and have, as you will see later in this chapter. Unfortunately, many of the victims of the Boxing Day tsunami did not have an understanding of how tsunamis work and could not recognise the warning signs. In some cases, this knowledge would have saved their lives.

Some waves are very, very big – not necessarily in terms of wave height, but in terms of overall size and the extent to which they can affect huge stretches of beaches at the same time. While dangerous tsunamis occur very rarely, there are other big waves that occur on a daily basis on every ocean beach. For example, most beachgoers are vaguely aware that the water in the ocean goes up and down *The waves smashed into beaches and coastlines in their path, reaching estimated heights of 30 metres ...* because of the tide, but may not understand why the time and the height of the tide changes every day, or why some beaches have massive tides and some almost none at all. What most people aren't aware of is that the tide is really a wave.

This chapter describes two types of tidal waves: the common ones that cause the gentle rise and fall of the waterline on our beaches every day and the rare catastrophic ones that are not really tidal waves at all, but tsunamis. It also describes the impacts that big waves created by severe storms can have on our beaches.

# TIDAL BORES ARE NOT BORING

Although the real tidal wave doesn't break on beaches, under exceptional circumstances it can break in the form of a reality-defying phenomenon called a tidal bore. Tidal bores are waves that travel upstream in coastal rivers as a result of the interaction between the outgoing river flow and the incoming ocean tide. They can appear as a single wave or a series, and break as spilling waves, but require very specific conditions. First, the tide range needs to be very high. Second, the river channel must be shallow and uniform at its mouth, becoming narrower and shallower upstream, creating a funnelling effect. Third, the speed of the tidal wave must be faster than the river flow.

Tidal bores only occur in about a hundred rivers around the world, including a few remote (and small) ones in northern Australia. They can reach heights of up to 9 metres, travel at speeds greater than 30 kilometres per hour and can extend tens to hundreds of kilometres inland. Bores such as the Silver Dragon in Zhejiang Province, China, and the Petitcodiac in New Brunswick, Canada, are renowned tourist attractions. Others have become famous as infrequent and bizarre surfing destinations. The River Severn bore in England provides rides of up to 10 kilometres, while the powerful and destructive Pororoca bore in Brazil's Amazon River has become the greatest challenge of all. Surfers can get rides of up to 40 minutes long, but they also have to deal with a churning mixture of muddy water, logs, alligators, piranhas and little candiru fish, which are particularly nasty for reasons best left to a google search!

# The real tidal wave

One of the most popular reality television series is *Survivor*. Early seasons typically involved dropping groups of strangers together on a tropical island to fend for themselves Robinson Crusoe-style. The intrepid survivors usually recognised the need to build a shelter, but invariably one group would build their hut on the beach, completely oblivious to the meandering line of flotsam and jetsam along the top of the beach otherwise known as the high tide mark. The hut would be built and they would then focus their attention on rubbing two sticks together to make fire. Suitably distracted, they never seemed to notice the waterline gradually moving up the beach until one night, much to their amazement, their hut is swamped by waves.

There are several take-home messages here. First, the casting process of *Survivor* clearly targets people who have no recollection of any necessary survival skills that were obvious from previous seasons of the show. Otherwise they would have figured out how to make fire in advance. Second, they didn't have a clue about how the tide works.

The tide is extremely important to beaches and beach users. Not only can it dramatically alter the appearance of a beach within hours, it can have a huge impact on the quality of waves and surf at your local break. Knowledge of the tides is also vital for visitors to places with large tides like the Broome region of Western Australia (Pic. 15), particularly when they tie up to a dock only to return six hours later to look down at a boat sitting on mud flats being jostled by crocodiles. Similarly, visiting the famous abbey at Mont-Saint-Michel in Normandy, France, at the wrong time can turn a leisurely stroll across some pretty tidal flats into a sudden sprint to outrun an incoming tide!

## What goes up must come down

Tides are waves because they can be described in the same way that waves breaking at the beach can. Every wave has a crest and a trough. So does the tide: high tide is the crest, low tide is the trough. They are distinctly different though. Tides have extremely long wavelengths and periods on the order of 6 to 12 hours. And they don't really break except in exceptional circumstances, in the form of tidal bores. The big difference, however, is that they are not formed by wind. To understand what creates tides and how they work, we need to turn our heads up and look towards the moon and the sun. It's all very cosmic.

All waves need a disturbing force to move the water, and in the case of tides, it's provided by the combined gravitational pull exerted by the moon and the sun on the water in the oceans, but mostly by the moon because it's closer. The key word here is 'oceans' because despite their size, large bodies of water such as the Mediterranean Sea and the Great Lakes don't experience any tide (well, maybe a centimetre or two). The reason for this is that relative to the overall surface area of the earth these bodies of water are mere blips and experience the same gravitational pull everywhere. Therefore, there is no displacement of water (disturbance) to create the tidal wave in the first place.

Imagine that the earth was all water. The moon's gravitational pull will create a bulge of water on the side closest to it. Another bulge is created on the opposite side of the earth because of the rotation of the earth. Ever heard of centrifugal force – the sort of thing that happens when you take a corner too fast in a car and your passengers' faces are pressed against the windows on one side only? It's the same sort of effect. So now the earth's water surface has two big bulges of water (high tides) and two flat bits (low tides) in between.

## The timing of the tides

Now pick up a set of tide tables from a local surf or fishing shop and look at the numbers. You will notice that the timing of high and low tide is approximately 50 minutes later each day. This is where Astronomy 101 comes in handy. We know that the earth rotates and one rotation takes 24 hours (a day). Assuming the moon stayed in the same place and we added a little island, with some nice beaches and a hammock, somebody lying in that hammock would watch two high tides and two low tides at the same time each day. The problem is, the moon revolves around the earth and each orbit takes 29 days (a lunar month). Hmmm.

This means that every time the earth completes one of its own 24-hour rotations, the moon and the ocean bulges have shifted in position a little bit and the earth has to rotate a bit longer to catch up with the bulges. As it turns out, it takes the earth about an extra 50 minutes each day to do this. If your beach experiences two high tides and two low tides a day, we say it has a *semi-diurnal tide* and this is very common on beaches around the world.

However, some beaches only experience one high and one low tide each day and this is called a *diurnal tide*. Why does this occur? As the earth spins on its axis, it also tilts a little so that the moon's gravitational force is shifted in position again. In some cases, a beach on the earth's surface will rotate under one bulge, but will miss the other bulge completely. The west and northern coasts of the Gulf of Mexico and parts of South-East Asia are examples of diurnal tidal environments. Some areas, like Torres Strait, between Australia and Papua New Guinea, experience mixed tides, which are a combination of diurnal and semi-diurnal tides – not surprisingly, the tides are all over the place!

*... the timing of high and low tide is approximately 50 minutes later each day.*

If this all sounds baffling, the good news is that the behaviour of the earth–moon system is well understood and totally predictable, which is why the timing of the tides can be predicted years in advance. All we need to do is pick up some tide tables and not worry about the mumbo jumbo!

## Tide range: springs and neaps

If you are particularly observant, you will notice that the tide at your beach comes up a little higher and goes out a little further each day ... and then reverses the process about a week later. The reason for this is that the *tidal range* varies from day to day. Tidal range is the vertical distance between high and low tide. In other words, it's the wave height of the tidal wave.

Let's assume your beach has a tide range of 1.8 metres. If you also happened to be 1.8 metres tall (six foot) and stood at the waterline at low tide and did not move, the water level would rise gradually and eventually stop atop your head at high tide (don't try this without a mask and snorkel). It would then take 6 hours to go back down to the level of your toes at low tide. In other words, the tide has gone up and down by 1.8 metres. Of course, there is yet another complicating factor

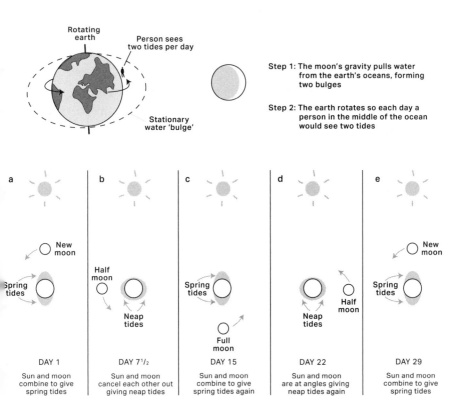

Step 1: The moon's gravity pulls water from the earth's oceans, forming two bulges

Step 2: The earth rotates so each day a person in the middle of the ocean would see two tides

| | | | | |
|---|---|---|---|---|
| **a** | **b** | **c** | **d** | **e** |
| New moon | Half moon | Spring tides | | New moon |
| Spring tides | Neap tides | Full moon | Neap tides / Half moon | Spring tides |
| DAY 1 | DAY 7½ | DAY 15 | DAY 22 | DAY 29 |
| Sun and moon combine to give spring tides | Sun and moon cancel each other out giving neap tides | Sun and moon combine to give spring tides again | Sun and moon are at angles giving neap tides again | Sun and moon combine to give spring tides |

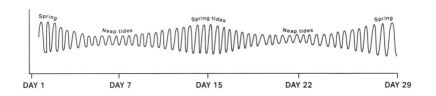

Tides are formed by the gravitational pull that the moon and the sun exert on the oceans of the earth. The height of the tide (tide range) changes every day and is related to the position of the moon and sun to each other and the earth. Larger spring tides and smaller neap tides each occur twice during a 29-day lunar cycle.

because the tide range just so happens to change from day to day, month to month and even year to year. Why? This is where the sun comes in.

Just like the moon, the sun also exerts a gravitational pull on the earth, but it's about half as strong because it's further away. Nevertheless, it's still strong enough to cause its own bulges in the earth's water surface. Since the moon revolves around the earth and the earth–moon system revolves around the sun, at some point the sun and the moon will line up so that their gravitational forces combine as shown in the diagram on page 63. This creates a bigger tidal bulge than normal, which results in higher high tides and lower low tides. Although these big tides have nothing to do with seasons, they are called *spring tides* and cause periods of higher tide ranges that last for several days. They also occur around a new moon and a full moon.

When the moon and the sun are at right angles to each other, they tend to cancel out and weaken their gravitational pull on the earth and the tidal bulges become smaller. This happens around the half-moon and causes smaller tide ranges called *neap tides*. Neap tides have high tides that aren't that high and low tides that aren't that low. So during every 29-day lunar cycle, beaches will have two periods of spring tides and two periods of neap tides. As shown on page 63, the tide range during each day of this cycle is different. What the diagram doesn't show is that the tide range from one lunar cycle to the next also changes. The reason for this is that sometimes the earth is further away from the moon and the sun, making the tide range smaller, and sometimes it is closer, making the gravitational pull stronger and creating extra large spring tides known as king tides.

# THE KING
# OF TIDES

There are a number of common myths involving the ocean and one is that every now and then beaches will get mysteriously swamped by 'king tides' and beachgoers are asked to watch out for hazardous surf conditions. It's a bad rap. A king tide is just a spring tide that has a bigger tide range than usual and is caused when the moon and sun are not only lined up, but are at their closest distance to earth. King tides are perfectly natural and totally predictable and occur twice each year, between December and February and between June and August.

Are they dangerous? Not really. King tides don't suddenly appear out of nowhere, drowning people on the beach. The rising tide is a gradual process and all that happens is that the spring high tide will come a little further up the beach and the low tide will go out a little farther. However, people should be wary of swimming in and around narrow river mouth entrances, particularly on coastlines with large tides, as king tides can enhance the strength of tidal currents. The biggest impact of king tides is the minor flooding of waterfront property built too close to the water for comfort. Beach erosion and coastal flooding can also be worse if large storms coincide with king high tides. Otherwise, don't worry about them!

# Tides around the world

The addition of the sun may explain why the tide range changes from day to day, but it doesn't explain why it changes from place to place. You'll find that the tide range at beaches around the world can be quite different, even beaches that are on other sides of the same island or peninsula. Because our hypothetical planet is made up only of water, it has one fundamental flaw: it ignores the landmasses in the middle of the oceans.

This is important because as the tidal waves (bulges) begin to move, they start bouncing off and refracting around things like continents, islands, reefs and the uneven sea floor. This causes the tidal waves to swirl around the middle of oceans in large circles that eventually get locked in place. The middles of these circles are called amphidromic points and the tidal range is zero. Just like a child's see-saw, the further away from the centre you go, the bigger the up and down movement is.

*... the tide range at beaches around the world can be quite different, even beaches that are on the other side of the same island ...*

So tide range on beaches increases the further they are away from these points. More advanced books on oceanography and marine science have maps showing the locations of the oceans' amphidromic points.

The tidal wave can also be squeezed by the shape of the coastline. In the case of long, straight open ocean coasts, like the southeast coast of Australia or the east coasts of North and South America, there isn't much squeezing going on, so the tide range is lower and much the same along the entire coast. However, head further north up the east coast of Australia and you eventually encounter the Great Barrier Reef. The tidal wave gets squeezed between the reef and the mainland and the tide range suddenly jumps from around 1.5 metres south of K'gari (formerly Fraser Island) up to 6 metres in places as you go north from K'gari towards Cairns. In general, if the tide range is less than 2 metres, we say that it is microtidal. If it's greater than 4 metres, it's macrotidal, and anything in between is called mesotidal.

## Tsunamis: waves of mass destruction

For most people the term 'tidal wave' conjures up an image of a single giant plunging wave that curls over and crashes on beaches, causing great loss of life and damage. In reality using the term 'tidal wave' to describe a tsunami is incorrect because tsunamis have nothing to do with tides and are not formed by the moon. 'Tsunami' is an old Japanese word combining 'tsu' for harbour and 'nami' for wave and was coined by fishermen who came back from fishing trips in perfectly calm wave conditions only to find massive destruction caused by large waves in their harbours and ports. To them, the wave that caused this damage must have been a 'harbour wave' because they had not noticed anything unusual while fishing out at sea.

Before 2004, very few people would have known much about tsunamis. The most famous example resulted from the 1883 eruption of the volcano Krakatoa in Java, Indonesia. Landslides and the collapse of the volcano created waves of up to 40 metres in height, wiping out nearby villages and killing over 40 000 people. Evidence of the tsunami was felt

# THE BIGGEST TIDES IN THE WORLD

Canada is famous for a number of things: ice hockey, doughnuts, big bears, a nice flag, Mike Myers, the more scenic Niagara Falls, being the only country ever to have hosted a summer Olympic games and not won a gold medal, and the biggest tides in the world. Once again, it's all about the geology. The Bay of Fundy separates the provinces of New Brunswick and Nova Scotia and has the perfect geometry to create a huge tidal range. It's funnel-shaped, being 290 kilometres long and 100 kilometres wide at its mouth, tapering and shallowing gradually along its length. This squeezes and amplifies the tidal range enormously so that the highest tide range at the end of the bay reaches an amazing 17 metres. That's a lot of water going up and down every day.

The importance of the coastline geometry is even more obvious when you consider that the tide range is only 4 metres at the mouth of the Bay and 2 metres on the nearby open ocean beaches. Other famous big tides around the world include Mont-Saint-Michel, the quasi-island and church abbey in France, which has the highest tide range in continental Europe (15 metres), and the 10-metre tides in parts of Western Australia, which cause the 'horizontal waterfalls' through a narrow rock chasm near Derby and the 'stairway to the moon' effect at Roebuck Bay in Broome, when the full moon reflects off the exposed mud flats at low tide.

as far away as the west coasts of North and South America. However, memories and lessons learned fade quickly. The catastrophic 2004 Boxing Day disaster (Pic. 16) changed all that and despite efforts at improving warning systems and public education, several deadly tsunamis have occurred since 2004, including the Tōhoku tsunami on 11 March 2011. An earthquake with a magnitude of 9.1 on the Richter scale, the largest ever recorded in Japan, generated a tsunami wave that reached the coast in approximately 30 minutes, travelling up to 10 kilometres inland, reaching estimated heights of 40 metres. Despite Japan being a world leader in tsunami preparedness, the tsunami resulted in at least 20 000 fatalities and a damage cost to Japan at an estimated US$235 billion. While there's nothing we can do to stop tsunamis from occurring, improving protective infrastructure, as well as recognition and response among people who live along tsunami prone coasts can only help save lives in future events.

## What causes tsunamis?

Imagine bumping into a small fish tank. The water in the tank will instantly slosh to the other side. In much the same way, tsunamis are generated by a disturbing force that suddenly pushes massive amounts of water. Earthquakes, underwater landslides on the slopes of continental shelves and the eruption of underwater volcanoes can all cause this displacement of water and start a tsunami. When a tsunami does form, it's not just one wave, but usually a series of waves having wave periods of several minutes to hours. It's similar to the effect of dropping a rock in a pond and watching a number of waves radiate outwards. Tide gauges in the Indian Ocean showed that the Boxing Day tsunami consisted of seven individual waves, with some being bigger than others. If you ever experience a tsunami and the first wave is not that big a deal, it's still a good idea to head for the hills.

Regardless of how tsunamis form, they are impossible to predict because the occurrence of earthquakes, volcanic eruptions and landslides are also impossible to foresee. We can only monitor a tsunami after it has formed. To do this we rely on a network of seismographs, which pick up earthquake signals, tide gauges, which monitor water levels, and ocean wave rider buoys, which can detect the presence of a tsunami wave. The good news is that once tsunamis form and start moving across the ocean, they behave like normal waves and it's possible to estimate their speed and direction. The bad news is that there's no stopping them and they are much bigger than normal waves in almost every way.

## Faster than a speeding jet

Tsunamis can have extremely long wavelengths of several hundred kilometres, which is why they can travel at speeds of up to 1000 kilometres per hour. That's faster than the cruising speed of most airliners! At the same time they have very small wave heights in deep water and are barely noticed by ships. Even the Boxing Day tsunami had a height of less than 1 metre in the middle of the ocean. Whereas normal wind and swell waves start to slow down, steepen and break when they get close to the beach, tsunamis are so big that they start to shoal when they interact with the continental shelf. If a coastline has a very wide continental shelf that also happens to be quite shallow, a tsunami wave will begin to shoal early and can build up to extreme wave heights by the time it reaches the shoreline. The opposite is true for islands

# THE CURIOUS CASE
# OF THE MALDIVES

The tiny island nation of the Maldives sits in the Indian Ocean, south of India and Sri Lanka, and is made up of 21 atolls, strung like a necklace, and studded with over 1200 coral reef islands. Think of gorgeous tropical cays, aquamarine lagoons and over-water bungalows perfect for honeymoons. That's the Maldives. They also have an average elevation of about 1 metre. It would be easy to assume that these tiny islands would have been wiped out by the 2004 tsunami because they were right in its path. But they weren't.

A total of 83 people were killed and of the 196 inhabited islands, 53 suffered severe damage and a third of the total population were affected. Of the 87 resort islands, 19 were severely damaged and closed down. It could have been a lot worse. While almost all the islands were overtopped by water, the flooding was relatively gentle and shallow with only minor damage to the islands and structures. In fact, many of the native villagers didn't realise a wave was involved – they thought their islands were sinking!

The Maldives were largely spared because they sit like pinheads in the middle of a deep ocean and are mostly surrounded by very narrow reef platforms. The tsunami was so big, it barely even registered the atolls. The islands that did experience significant damage were all surrounded by wide reef platforms that allowed the tsunami waves to slow down, shoal and build up some dangerous wave heights.

such as the Maldives that have no continental shelf. Although tsunamis slow down to about 50 kilometres per hour when they hit the beach, that's still a lot faster than people can run.

## Tsunamis are not plunging waves!

The most famous image associated with a tsunami is a woodblock print, *The Great Wave off Kanagawa*, created in the 1820s by Japanese artist Hokusai, which shows a plunging wave dwarfing fishing boats and Mount Fuji in the background. It's an impressive image that is somewhat misleading because the artist was depicting storm waves in the ocean, not a tsunami. Furthermore, tsunamis are not plunging waves. Nevertheless, most people still think of them as towering walls of water rearing up to plunge down and smash everything in front of it. In some ways, it would be better if it did as this would reduce the damage. One of the reasons tsunamis cause so much death and destruction is because they are surging waves (Pic. 16).

Because they surge, tsunamis can reach considerable distances inland, especially on flat coastlines. While the uprush of the tsunami can knock over many objects in its path, the return backwash is just as lethal because it is full of floating debris that can knock out and trap people who are trying to remain afloat.

Fortunately for people at the beach, tsunamis sometimes come with their own warning sign that can give you some valuable time to reach the safety of higher ground. Eyewitness reports often describe the water at the beach mysteriously receding considerable distances offshore before the arrival of a tsunami. In some instances, entire bays have been emptied of water in minutes, leaving schools of surprised fish flopping about. This happens when the trough of the tsunami hits the coast first. Imagine lying on a raft in small waves close to the beach. Water motion in the trough will pull you offshore before the crest of the wave pushes you towards the beach. It's the same with tsunamis except the water is pulled much greater distances offshore because their troughs are kilometres long.

Tragically, many people do not recognise the danger of this phenomenon and instead become curious, innocently wandering further down the exposed beach simply to have a look. This can be a fatal mistake since the crest of the tsunami will soon arrive with great force and speed. There is a famous sequence of photos taken at Railay Beach near Krabi in Thailand during the 2004 tsunami

*... tsunamis sometimes come with their own warning sign that can give you some valuable time to reach the safety of higher ground.*

showing a family doing exactly this. Amazingly, they survived while tragically others didn't. If you ever do see water suddenly getting sucked out to sea, it's definitely not normal and the appropriate response is to turn and run. Preferably up a hill!

## Should you be worried?

Every ocean coastline on the planet is susceptible to tsunamis, but they are not something you should lose sleep over, even if you live by the beach. Most tsunamis are tiny in height and are barely noticeable. Damaging tsunamis are extremely rare, although the risk is certainly higher in some places. About 80 per cent of all recorded tsunamis have been in the Pacific Ocean, largely due to the extensive line of seismic and volcanic activity known as the 'Pacific Ring of Fire'. Hawai'i and Japan are exposed to large expanses of the Pacific Ocean and tend to get hit from all directions, so it's no wonder they have been key leaders in establishing tsunami warning systems and defences. While severe tsunamis may destroy a beach by stripping all the sand away, and indeed may reconfigure the coastline, beaches are remarkably resilient and the sand will quickly come back under normal wave conditions. I visited Railay Beach a year after the tsunami and the beaches were back to normal. Most visitors didn't even know the tsunami had impacted there, although if they had explored a little bit behind the beach, they would have found old huts and boats washed up in the forest.

Geology also plays an important role in the risk of a tsunami at a beach. For example, the Atlantic coasts of North and South America are relatively free from seismic activity and are not prone to tsunamis. Conversely, New Zealand is an active seismic area (they don't call them the Shaky Isles for nothing!) and has experienced numerous tsunamis, many caused by its own earthquakes. The southeast coast of Australia, including Sydney, should be in the direct firing line of numerous tsunamis but thanks to a number of factors, including its narrow and steep continental shelf and the bathymetry of the Tasman Sea between Australia and New Zealand, the risk of a severe tsunami is very low. Unless, of course, a huge meteorite falls into the Tasman causing a gigantic megatsunami 100 metres in height. Don't laugh, some people think this has happened before and who knows, maybe it did!

# Batten down the hatches: hurricanes and cyclones

In recent years, the world has witnessed two historically damaging ocean storms that in combination caused as much, if not more, damage than the Boxing Day tsunami. In late August 2005, Hurricane Katrina moved across the Caribbean Sea and the Gulf of Mexico, reaching peak wind speeds of 280 kilometres per hour and with an extremely low barometric pressure of 902 millibars, creating a maximum storm surge of 7.6 metres. This resulted in 1836 fatalities across the region and flooded 80 per cent of New Orleans, Louisiana. The total cost of Katrina was an estimated US$108 billion, and it is still the costliest hurricane in history.

In May 2008, Cyclone Nargis developed over the Bay of Bengal in South-East Asia and made landfall over low-lying regions of the Irrawaddy River delta in Myanmar – it has been described as Asia's 'perfect storm'. A storm surge of 3.6 metres resulted in flooding more than 40 kilometres inland, 146000 fatalities, 2 million people left homeless and damage estimated at US$10 billion. It is still one of the most deadly cyclones in history

While both Katrina and Nargis were physically very similar, it is interesting to note the difference in their impacts. Katrina caused massive economic damage, but relatively little loss of life, whereas the opposite was true of Nargis. The same is true when comparing the 2004 Boxing Day tsunami in Sumatra and the 2011 Tōhoku tsunami in Japan. Such is the difference in impact of natural hazards between developed and less-developed countries.

It is amazing, given the impacts of these storms, that humans have an ability to forget and become complacent about natural hazards. Have you ever watched TV news reporters standing on beaches being buffeted by cyclone winds, warning people to evacuate, while surfers run past them trying to catch some waves? Even after catastrophic storms we eventually rebuild, move back in and hope for the best. Beaches often get wiped out during severe storms due to massive erosion and although the sand will eventually come back on its own, we often speed the process along by nourishing the beaches at great cost so that the tourists will return. Unfortunately, it's almost inevitable that major disasters like Hurricane Katrina and Cyclone Nargis will happen again.

## Types of storms

Storms come in all shapes and sizes, but one thing they all have in common is strong winds, and with strong winds come big waves. Smaller storms – such as wind squalls, thunderstorms and the arrival of weather fronts – can cause messy surf conditions, but are rarely more than a temporary nuisance. Monsoons in the tropics can cause rougher wave conditions and can shift sand around on beaches, but they are not usually damaging except during cases of severe coastal lowland flooding. Larger storms – such as tropical cyclones, hurricanes and typhoons – are a whole different ballgame, but despite their different names, they are really the same thing.

They are all massive storm systems that revolve around a central and small region of low atmospheric pressure, often called the 'eye' of the storm. In the southern hemisphere, these storms are called tropical cyclones and rotate in a clockwise direction while moving from west to east through the tropical waters off Indonesia, northern Australia and the South Pacific. The east coast of Australia can also be affected by similar storms that form in the middle of the Tasman Sea and are called East Coast Cyclones. In the northern hemisphere, cyclones rotate in an anti-clockwise direction and are called hurricanes. A typhoon is just the name given to hurricanes that occur in the northwest Pacific Ocean and originates from the Chinese term 'tai-fung', meaning 'great wind'.

Most cyclones tend to form in summer months as they gain energy from the evaporation of warmer ocean waters. All cyclones can move relatively quickly and can pack winds of over 200 kilometres per hour while taking very unpredictable paths. The fact that they are often given cute names such as Hugo, Fifi and Winston belies the fact that they can be extremely deadly.

The East Coast Cyclones that impact the populated New South Wales coast of Australia are a little different. They are not given names, don't have wind speeds as strong as tropical cyclones and don't last as long, but can still cause fatalities, largely due to flooding associated with their extreme rainfall. They also cause significant beach erosion due to waves that can reach over 10 metres in height. Not surprisingly, they often result in severe damage to coastal properties and infrastructure. Some famous examples of East Coast Cyclones include the June 2007 Pasha Bulker storm where a freighter ship of the same name was washed up on Nobby's Beach in Newcastle and in June 2016, a section of houses, including a pool, ended up on the beach along Sydney's Narrabeen-Collaroy beach. However, regardless of what type of severe storm is occurring, the real problem isn't the wind and the waves, it's the storm surge.

# Storm surge

Storm surge is the rise in water level at a shoreline above the normal water level due to a combination of extremely low air pressure and strong onshore winds (Pic. 17). Cyclones tick both these boxes. Imagine someone standing on your chest. That's a lot of pressure and your chest will feel like collapsing. Once they step off, the pressure is reduced and your chest will expand again. Cyclones have such extreme low pressure that the water level in the ocean rises. Now combine that with incredibly strong winds that are creating nasty waves and literally blowing water towards the beach and suddenly you've got a storm surge and a big, big problem.

A storm surge can also be affected by the shape of the seabed and, just like a tsunami, is amplified on wide, flat continental shelves. It can also be squeezed higher by coastline geometry such as river mouths, tidal inlets and bays.

*... because they can last for longer periods of time, they are often more destructive than tsunamis.*

Of course, the worst-case scenario is a storm surge coinciding with a spring high tide. Not only will the beach and dunes be eroded under these conditions, there will usually be extensive inland flooding, particularly in estuaries, barrier islands and other low-lying environments. Low-lying coastal developments usually take a beating and can often be wiped out completely.

Along relatively flat coastal margins, there is very little to stop storm surges reaching considerable distances inland and because they can last for longer periods of time, they are often more destructive than tsunamis. Nowadays, at least for developed countries, there is much more warning time available before the arrival of extreme storms, which aids in evacuation plans and significantly reduces the potential loss of life. Technology helps as well. Both England and the Netherlands have built highly engineered storm surge barriers that basically clamp shut during storms, holding back the storm surge and stopping inland flooding. How times have changed since the story was told of a little Dutch boy who saved his town by plugging a hole in a dyke with his finger.

## Storms and climate change

The climate is changing, there's no doubt about it, and the obvious way that climate change will impact our coasts and beaches is through sea level rise, which was discussed in chapter 1. But how will climate change affect the magnitude and frequency of severe storms? In other words, what will happen if cyclones become stronger and occur more often? Well, it wouldn't be good. Let's use the East Coast Cyclones that impact the coast of New South Wales as an example. At the moment it seems like we get a really bad one about every 10 years on average. Despite the serious erosion that occurs, most beaches will recover within six months to a year. Great! But what happens if one of those storms starts happening every year? There won't be much opportunity for beach recovery and the coastal erosion problems will get even worse. The problem is that predicting the impacts of climate change on storms in the future is difficult and we just don't know for sure what is going to happen.

What we do know is that there's a big connection between what's going on in the oceans (think large currents and water temperatures) and what's going on in the atmosphere (think winds and rain). You may have heard of the climate phenomenon called the El Niño–Southern Oscillation (ENSO), which refers to a flipping of El Niño and La Niña conditions in the Pacific Ocean. Let me explain by again using Australia's New South Wales coast as an example and, conveniently, skipping a lot of background details. During an El Niño climate phase, the trade winds that blow from east to west across the Pacific start to weaken and surface waters in the central and eastern Pacific warm up, while those closer to Australia become a little cooler. For New South Wales this causes the weather to be warmer with fewer storms and less rainfall. El Niño phases can last for years and while long ones can unfortunately lead to droughts and bushfires, beaches love them because with fewer storms and gentler wave conditions, the sand builds up to create nice healthy beaches.

An El Niño phase won't last forever and will eventually transition into a bit of a neutral phase until the trade winds get stronger and a La Niña phase begins. In a La Niña, the water temperature close to Australia gets warmer and leads to more weather instability with more rain and storms. While it helps put out the fires, the increase in storms creates bigger wave conditions, resulting in eroding beaches. As I write this, the New South Wales coast is just starting to come out of a rare back-to-back-to-back triple La Niña and our beaches are a mess. Things haven't been this bad since the mid-1970s when, funnily enough, we had our last triple La Niña. The good news is that the beaches will eventually fully recover when the next El Niño kicks in.

While ENSO cycles take place on the order of years, there are other climate cycles that have been identified that affect our weather, waves and beaches over longer periods, such as the Pacific Decadal Oscillation and the Indian Ocean Dipole, which are ocean-atmospheric phenomena that can change weather and ocean conditions over decades. So will climate change influence these cycles and therefore the behaviour of our beaches and coasts? Absolutely, which is why climate change science is so important in order to improve our understanding of the complex relationships between oceans and climate and how our beaches will respond in the future.

# The bottom line

★ Tides and tsunamis are waves and have many of the same characteristics and behaviour as normal waves at the beach.

★ Tides are created by the gravitational pull on the water in the oceans by the moon and the sun.

★ Most beaches have two high tides and two low tides a day, but the timing of them is a bit later every day.

★ Tide range is the vertical distance between high tide and low tide and varies over time and from place to place.

★ Spring tides are times of bigger tides and neap tides are when smaller tides occur.

★ Tsunamis are extremely long and fast waves formed by earthquakes, landslides and volcanic eruptions that displace water in the oceans.

★ Most tsunamis that reach beaches break as surging waves.

★ Catastrophic tsunamis are rare and are impossible to predict.

★ Storm surge is caused during large storms such as cyclones or hurricanes and can be more damaging than tsunamis.

★ When you hear evacuation warnings for tsunamis and hurricanes, head inland or to higher ground as fast as possible. Do not stick around to watch!

# 4

# White is nice, green is mean

## Rips and other currents

Drifting in a rip current about 300 metres offshore (and counting) at Muriwai Beach, one of Auckland's high-energy west coast beaches, I started to think of the common advice given to people on what to do when caught in a rip current. 'Just swim to the side' came to mind. Yes, that was good advice, but possibly not so good given that the rip current was so wide I couldn't actually see either side. I suppose I could have taken a shot at it Olympic swimmer style, but I reasoned that by the time I made it to the side, if I made it at all, I'd be half way to Sydney so what was the point? Another stellar piece of advice was, 'Don't worry, the rip current will eventually bring you back to the beach'. Would it? Really?

I wasn't convinced. It seemed pretty doubtful to me, not being able to see the beach I had been standing on only minutes before. All I could see were walls of water. Finally I reminded myself that I should heed the advice, 'Relax and don't panic' and was amazed that this provided me with absolutely no comfort at all. All I really wanted to do was pee in my wetsuit. I wasn't sure if this was a sign of the extreme relaxation I was feeling because of my understanding on how to behave properly in a rip current or because I was on the verge of panicking.

Fortunately for me, I was floating as part of an experiment to measure the speed and trajectory of really big rip currents. Not too far away was an inflatable rescue boat with some very capable lifeguards aboard. By the time they pulled me out, I had travelled almost 300 metres along the beach in a channel feeding the rip current, which then took me almost 400 metres offshore at speeds approaching those of swimming world records. It was quite a ride, but if the boat hadn't been there, I would have been in big trouble. Rip currents are responsible for hundreds of drownings and tens of thousands of lifeguard rescues on beaches around the world every year. Folks, a lot of our beaches have a rip current problem.

*... understanding how currents work and how to spot them may just save your life.*

So what are these mysterious rip currents and why do they move so much water (and people) away from our beaches? Rip currents, or 'rips', are currents and just like river currents, they move water from one place to another, usually in one direction. So think of rips as rivers of the sea. Currents. Hmmm. Now there's a term that tends to make the average person's eyes glaze over. It shouldn't though because water moves all over the place on many beaches and understanding how currents work and how to spot them may just save your life.

Whether you are aware of it or not, if you've ever swum at a beach with breaking waves, you've experienced some sort

# BEWARE OF THE 'UNDERTOAD'

As a kid I was always told to watch out for the undertow when we went to the beach. Undertow was something in the ocean that sucked you under the water, held you there and then tried to drown you. All the parents seemed to know about it, so it must have been important and it sounded terrifying. Some kids even thought there was an 'undertoad' in the water ready to pull them down and eat them. Scary stuff.

The problem with this theory is that there is no such thing as an undertow. Contrary to popular belief, there is nothing in the ocean that will pull you under and keep you there. We have oxygen in our lungs and a natural tendency to float (buoyancy), so even if a wave crashes over you and pushes you to the bottom, you will eventually pop back up to the surface. Unfortunately rip currents are still called 'undertow' all the time. This is unfortunate because this misunderstanding leads to panic. Rip currents just take you for a ride. Rips are also commonly referred to as 'rip tides', which sounds nice and nautical, but is also misleading because a rip current is not a tide. Tides are slow shifts in water level over hours, whereas rips are strong currents. So the correct term to use really is 'rip current'. Or just 'rips' for short.

of current in action. The purpose of this chapter is to describe the different types of currents found along beaches and explain how and where they form and if they are dangerous to swimmers. It's also important to remember that currents don't just move water and swimmers around, they also move sand and play a huge role in determining what your beach will look like on a day-to-day basis.

# Rip currents

## Don't get sucked in by the rip

In 1902, Mr William Gocher, a newspaper editor probably trying to drum up some news, defied the ban on daytime swimming and went for a dip at Sydney's Manly Beach and was immediately arrested. Although the motivation for his actions remains unknown, it is often documented that this 'hero of the surf' opened the floodgates for Sydneysiders to start swimming en masse during the daytime, whereupon many drowned – mostly in rips. In response, the iconic Australian surf life saving movement was born and the first Surf Life Saving Club was established in 1907 at Sydney's Bondi Beach – or Bronte Beach, depending on who you talk to. It's a touchy subject. Since then, lifeguards and lifesavers around the world have made an awful lot of rescues and probably saved tens of thousands of lives from the perils of rip currents.

It could be argued that, being a poor swimmer, perhaps it would have been better if, instead of being dragged off by the police, Mr Gocher had been dragged off by the notorious 'Manly Escalator' rip, thus becoming the first high-profile rip current victim in Australia. Maybe then, from the word go, rip currents would have been recognised as the biggest hazard

## THE RIP CURRENT SURVIVAL
GUIDE | The best way to survive a rip current is not to get
in one in the first place. Make sure you always try and swim on a
beach that is patrolled by lifeguards. In Australia, New Zealand,
the United Kingdom and some other countries, the message is
to swim between the red and yellow flags, which denote safer
and supervised swimming areas on the beach. Other countries
just have lifeguards, so try and swim near them! If you are on a
beach with no lifeguards and you are not a good swimmer, make
sure you don't go in beyond waist depth and, even if you are a
good swimmer, learn how to spot rips so you can avoid them. If
you do find yourself drifting quickly offshore however, here is
some good advice.

**1. RELAX |** It's easier said than done when the beach is
disappearing rapidly, but if you get caught in a rip current, try to
remain calm and avoid panicking. Remember that a rip will not
pull you under the water, it won't take you to the other side of the
ocean, or into shark-infested waters, and there's a good chance
it will circulate you back into shallow water where you can stand
up safe and sound. It's just taking you for a ride. People ask me
all the time what it feels like being caught in a rip, but it doesn't
feel like anything, because you are just going with the flow. It's
panic that gets people in trouble and in the words of Bondi's
chief lifeguard Bruce 'Hoppo' Hopkins, 'Rips don't drown people,
people drown in rips'.

USING THE BEACH

**2. STAY AFLOAT AND SIGNAL FOR HELP |** The best way to try and relax if caught in a rip is to stay afloat. You can do this by either lying on your back and slowly moving your legs and arms, or by treading water. Either way, you are conserving your energy and giving yourself time to think about what you should do next … which is to signal for help! Don't be shy, if there are lifeguards around, raise your arm and wave for attention. If there are surfers around, call out for help. Surfers have a nice surfboard to hold onto and they do just as many rescues as lifeguards.

**3. DON'T SWIM AGAINST THE RIP |** Most people's natural reaction when stuck in a rip is to swim against it back to the beach. Don't. Even small rips can flow faster than the average person can swim. You may find yourself going nowhere or backwards and the harder you try and swim, the more tired you'll become and this will lead to panic.

**4. SWIMMING OUT OF THE RIP DOESN'T ALWAYS WORK |** Probably the most common advice that is still around today is 'to swim parallel to the beach' to escape the rip. This was based on the idea that rips flowed straight offshore and because they aren't that wide, you only need to swim a little way either side before you're out. Easy right? Well, in fact it's not so easy.

Rips don't always flow straight out from the beach, they can flow at angles as well as bend and curve. They can also be anywhere from 5 to 50 metres wide. We've done lots of experiments putting people (mostly volunteer students!) in rips and asking them to swim parallel to escape the rip (Pic. 18). Sometimes it was easy and took less than a minute, but other times people ended up swimming against the rip current flow and didn't make it and just tired themselves out.

Instead, if you are an experienced and strong ocean swimmer and you want to swim out of a rip, your best bet it to just look for whitewater and breaking waves and head for that. Whitewater is good because it means it's shallower and you may be able to stand up. Whitewater will also help bring you back to shore. When it comes to most rips, 'white is nice, green is mean'!

**5. WORST-CASE SCENARIO |** I often get asked something along the lines of, 'What should I do if I'm caught in a rip on a beach where there aren't any lifeguards or anyone else to help me and the rip takes me way offshore?' Good question. Your only real option is to swim a long way up or down the beach to get away from the rip and then swim back to shore where there are waves breaking. That's a long swim. An even better question would be 'Why were you even swimming there in the first place?' If you are on a beach and there are a lot of breaking waves and there's no lifeguards or people around and you don't know how to spot a rip then DON'T go swimming. If in doubt, don't go out. It's as simple as that.

on our beaches and would have generated the interest and attention among both the public and media that they deserve.

Shark attack? Front page news. Rip drowning? Barely rates a mention. There often seems to be a dangerous degree of complacency and acceptance about rips. In 1967, Harold Holt, then Australian Prime Minister, went for a swim just before Christmas at Cheviot Beach near Portsea, in Victoria, only to drown in a rip. Nevertheless, some people still believe the conspiracy theory that he really secretly rendezvoused with a Chinese submarine. Rips just don't get any respect.

They are also called a lot of different names, most of which are incorrect or misleading. This creates a lot of confusion. The term 'rip current' was coined by oceanographers working in La Jolla, California, in the 1920s. The origins of why 'rip' was chosen remain unclear, but it's probably because most rips look like a tear, or gap, through the surf. If you want a nice, simple definition of what a rip is, here goes: rips are strong, narrow 'rivers of the sea' that flow from close to the shoreline offshore to the extent of breaking waves, and sometimes beyond, at speeds sometimes faster than the average person can swim. The question then becomes: why?

*... rips are strong, narrow 'rivers of the sea' that flow from close to the shoreline offshore to the extent of breaking waves ...*

## No breaking waves = no rips

Not all beaches have rips. Some never have them, some only have them occasionally, while others have them almost all the time. They can occur on ocean beaches, inland seas and large lakes. Basically any time a beach experiences waves breaking over a wide area, it will have rips. If the beach also has sand-bars and channels that shift around, even better, because rips like to sit in deeper channels squeezed in between shallower sandbars. The key ingredient though is breaking waves. If there are no breaking waves, there won't be any rips.

There are three important things about breaking waves that are critical to understanding why rip currents exist. First, the whitewater you see after a wave breaks is water that is being physically moved towards the beach. Water can't pile up on the beach all day or else we'd all be under water, so somehow it has to head back offshore to keep everything balanced. Second, although you can't see it with the naked eye, the water level rises a little where waves are breaking. What this means is that water levels are always a little bit higher where there are lots of breaking waves compared to areas where there is less wave breaking, or no waves breaking at all. The same is true where bigger waves are breaking compared to where smaller waves are breaking. In both cases, water will start to flow from areas of higher water levels to areas of lower water levels. In other words, the water is basically flowing downhill. Third, waves don't break the same way everywhere and all the time. There are all sorts of things along the beach, like sandbars, headlands and rock reefs that cause waves to break a lot in some places and not in others. It's this variability of wave breaking over space and time that really drives rip current flow. It's also what causes different types of rips to form.

Scientists have been measuring how rips flow for years. The general consensus is that the bigger the breaking waves, the bigger and faster the rips will be. We also know that on most beaches, rips flow fastest several hours around low tide because the shallower depths make wave breaking more intense. Most rips will slow down and some may even stop completely at high tide because there just aren't enough breaking waves to create rip current flow. Lifeguards often post the tide times on a notice board and it pays to check them because, in terms of rips, it's much safer to swim around high tide. How fast do rip currents flow? A typical rip current will have an average flow speed of about half a metre per second. To put that in perspective, an adult standing in waist-deep water in a rip that strong would find it hard to stay in place!

## Rip dynamics

If you ever stumble upon a textbook on coastal processes, it will no doubt have a diagram that shows an ideal rip current system that behaves like this: waves bring water towards the shore where it then starts moving along the beach as a feeder current. If two feeder currents flowing in opposite directions meet, they combine and turn straight offshore, forming a narrow and fast-flowing rip neck. This rip neck extends through the surf zone and well past the line of breaking waves. Then it starts to slow down, spreading out into a mushroom-shaped rip head, and eventually dies out. The water is then free to be carried back towards the beach by waves on either side of the rip head.

Does this happen in real life? Well, yes and no. There are plenty of rip currents that behave exactly like this. There are also plenty that don't. Some rips flow at angles offshore and some have only one feeder, or none at all. Often the water enters the rip by draining off the sides of adjacent sandbars. Some rips carry water (and swimmers) way past the line of breakers and some don't.

In 2009 I was lucky enough to spend a few months visiting Jamie MacMahan in Monterey, California. Jamie is

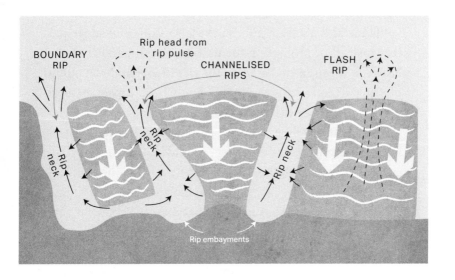

How different types of rip currents work. Breaking waves bring water to the beach and rips help take it back out. Rips flow at different angles to the beach, carrying swimmers various distances offshore.

a rip current guru who had just conducted a bunch of field experiments putting drifters in rip currents with Global Positioning Systems (GPS) attached (Pic. 18). Jamie and his team would carry a bunch of drifters into rip currents and just let them go and let them do their thing. What he found changed the whole way we think about rip flow behaviour. Whereas the traditional understanding of rip currents suggested that rips flow way beyond the line of breaking waves, Jamie found that most of his drifters actually flowed in circles without leaving the surf zone. They would flow along the beach in a feeder channel, turn offshore in the rip-neck and then instead of heading offshore, they would circulate back across the sandbars and towards the shore with breaking waves. In many cases they would then end up back in the rip again like a merry-go-round.

Jamie and his team published their results and suggested that instead of promoting the message 'swim parallel to the beach' to escape a rip based on the traditional understanding of rip currents, maybe people should just conserve their energy and float because there's a good chance that in a few minutes, the rip flow will circulate them back into shallow water where they can stand up. Sounds good! If every rip current flowed in circles, it would be fantastic! The problem is, they don't. Jamie's experiments and many others, including some of our own on Australian beaches, have shown that every now and then rip currents have a very bad habit of flowing a long way offshore beyond the surf zone before they stop. In this case, the only thing bringing you back to the beach would be a very long swim.

## Rips behaving badly

For years I've been putting harmless purple dye into rip currents to demonstrate how rip currents behave. The dye really helps bring the rip current to life. Sometimes the dye moves out quickly, sometimes slowly, sometimes it circulates back to the beach (Pic. 19) and sometimes it flows well beyond the surf zone (Pic. 20). That's the problem with rip currents – you just never know what they're going to do!

One of the reasons for this is because rip current flow isn't steady, in other words, they don't flow at the same speed all the time. In fact, rip current flow is quite unsteady and they all have a nasty habit of suddenly increasing their speed, sometimes doubling it, in a matter of seconds. These sudden pulses have been measured at more than 2 metres per second, which is similar to the world record for swimming the 100 metres freestyle. A lot of Olympic swimmers trying to swim against a rip pulse would end up going nowhere, or backwards. The average swimmer would have no chance.

We do understand why pulses in rip flow occur and, once again, it's all about breaking waves. When a wave set, which is a group of large waves, has rolled in and broken, the extra amount of water brought in by the larger breaking waves piles up near the beach and basically pumps the rip as it makes its way back offshore. Rip pulses can be dramatic and are the main reason why rip current flow sometimes exits the surf zone (Pic. 20). Not surprisingly, they are also one of the main causes of rescues, and often mass rescues, as they can suddenly sweep swimmers from the relative safety of shallower water into the main part of the rip and offshore. All rips, big or small, can pulse so it's another reason to pay attention to what the waves are doing when you are swimming. When a big wave set comes in, it's a good idea to get back into shallow water.

## Types of rips

It cannot be said enough that the best way to avoid rips is to learn how to spot them. Unfortunately this isn't easy to the untrained eye. One of the first things I did when I arrived in Australia from Canada in 1992 was to head to Sydney's Bronte Beach for a swim, where Ian Turner, a graduate student at the University of Sydney (and now a good friend and colleague), pointed out the rip current I needed to avoid. The only problem was, I couldn't see it. Even when he kept pointing it out to me, I just couldn't see it. What really disturbed me was that I had studied rip currents as an undergraduate student!

Aside from feeling rather silly, I was intrigued. If I studied rips out of a textbook and still couldn't recognise one in real life, what chance did everyone else have? Not much apparently, as studies have shown that the majority of beachgoers cannot spot a rip current to save their life. Part of the reason why is that no one is teaching them how to. Another problem is that there are different types of rip currents that not only look different, but behave differently. So if you really want to know how to spot a rip, you also need to know something about the types of rip currents that are out there.

During one of my university sabbaticals I worked with Bruno Castelle at the University of Bordeaux in France, who is one of the best coastal and rip current scientists on the planet. We decided to work on the ultimate classification of rip current types and gathered together some other fantastic rip current scientists, Tim Scott from Plymouth in England and Jak McCarroll from Australia, to help us. It was mere coincidence that we were all friends, all liked to surf and, well, there are some beautiful surf beaches near Bordeaux and the wine is plentiful and cheap, and we did have a late night trying to play Vance Joy's 'Riptide' (bad term!) badly on a variety of musical instruments, but I digress. Led by Bruno we came up with the following types that I have grossly simplified here, but just search for 'Rip current types, circulation and hazard' online and you'll find the full study in all its glory.

## Channelised rips

The most common type of rip in Australia and globally is one that sits in a deep channel snuggled between shallow sandbars that extends through the surf zone (Pic. 21). These channelised, or 'channel' rips occur on any beach with breaking waves across a surf zone characterised by variable patterns of

# MASS RESCUES AND THE MYTH OF COLLAPSING SANDBARS

Sunday, 6 February 1938, at Sydney's famous Bondi Beach was a beautiful summer's day with many of the 30 000 people at the beach enjoying the surf. Then three large waves approached the shore and broke. Soon after, swimmers suddenly found themselves being dragged into a deeper channel of water and out to sea. Approximately 60 surf lifesavers conducted a mass rescue during the ensuing hysteria and panic. In the next 30 minutes, 250 bathers required assistance, of which 35 were rescued unconscious and revived, while tragically five drowned.

Today this mass rescue is still often reported as being caused by a collapsing sandbar, which is unfortunate as sandbars simply do not collapse! The report of three large waves and a deep channel tells an interesting story though. The deep channel was probably a rip channel and the larger waves likely represented a wave set coming in. As the waves broke, the water level would have risen and swimmers standing on the sandbars near the rip would have lost their footing and floated into the rip. When the rip pulsed, they would have been taken quickly offshore. Stories of collapsing sandbars are not uncommon, but it's almost guaranteed that in each case it was a rip pulse. However, although not proven, there is some evidence to suggest that the waves that came in at Bondi that day were not part of a wave set, but those of a small tsunami!

sandbars. This variability between shallow and deeper areas also creates variability in areas of wave breaking, which is what drives the rip current flow.

Channelised rips can stay in the same place for days, weeks and even months if wave conditions don't change much. They also occur when waves are relatively small and although they can eventually be destroyed by the action of big waves during storms, they can quickly form again afterwards. They are usually between 5 and 30 metres wide and once they form, you don't just get one, but a whole bunch of them spaced along the beach at semi-regular intervals. On the south-east coast of Australia they tend to occur every 150 to 200 metres along the beach. They've often been given different names, such as fixed rips, beach rips and bar-gap rips, but 'channelised' makes the most sense, because that's what they are.

Channelised rips are the easiest to spot as they appear as dark gaps of seemingly calmer water between areas of breaking waves and whitewater. A good way to remember how to spot them is to use the saying 'white is nice, green is mean' (Pic. 21)! However, despite the fact that they are easier to spot and occur in small surf conditions, lifeguards will tell you that these are the most dangerous types of rip currents in terms of drowning because they exist on beaches on beautiful days when more people visit the beach and are likely to go in the water because conditions seem so inviting. Unfortunately because channelised rips appear to be calmer areas, many inexperienced beach users think they are also the safest place to swim. So in they go … and off they go!

**Pic. 1.** There are lots of reasons we love the beach, including watching the planes land at Princess Juliana International Airport on the Caribbean island of Sint Maarten.

*Wikimedia Commons user Lawrence Lansing*

**Pic. 2.** Geologic control at its best in Sydney's Eastern Suburbs beaches. From top to bottom: Bondi, Tamarama, Bronte, Clovelly (hidden), Gordons Bay and Coogee beaches.

*Rob Brander*

Pic. 3. There's more to sand than meets the eye. A pinch of sand from Coalcliff Beach in New South Wales under a microscope reveals a kaleidoscope of colours and shapes.

*Rob Brander*

Pic. 4. Fire and ice. Diamond Beach near Jökulsárlón in Iceland has both volcanic black sand and fragments of glacial icebergs.

*Rob Brander*

**Pic. 5.** The beautiful carbonate beach on Lady Elliot Island on the Great Barrier Reef is nice and white, but also made up of broken bits of coral. Not suitable for bare feet!

*Rob Brander*

**Pic. 6.** Yes that's a surfer looking very small on a wave in Nazaré, Portugal, that is at least 20 metres high. Nazaré has some of the biggest waves in the world.

*Shutterstock, photographer aleksey snezhinskij*

Pic. 7. Whitecapping and choppy wind waves forming in Sydney Harbour during a 'southerly buster'. Only minutes before the water was flat calm.

*Rob Brander*

Pic. 8. Clean lines of swell with long wavelengths and periods rolling in from a distant storm. No wonder they call it corduroy.

*Rob Brander*

Pic. 9. Beautiful lines of swell refracting around the headland at Cape Byron onto The Pass in northern New South Wales.

*Rob Brander*

**Pic. 10.** A classic plunging wave breaking over a reef in Sumatra's Mentawai Islands, creating a perfect barrel for surfing.

*Tim Scott*

**Pic. 11.** Spilling waves break on gentle beach slopes when the crest steepens and crumbles, cascading the water towards the beach. Perfect for learning how to surf.

*Rob Brander*

**Pic. 12.** A sneaky surging wave bulging up and about to break while another section has already broken and rushed up the beach at Sarge Bay in Western Australia.

*Rob Brander*

**Pic. 13.** Ouch. Steep beaches can create some nasty shorebreaks known as 'dumping waves'. Not a good bodysurfing wave.

*Rob Brander*

**Pic. 14.** A reflected wave from the cliff meets an incoming wave off the Caribbean island of Martinique. Boogie boarders love this sort of thing.

*Wikimedia Commons user rachel_thecat*

**Pic. 15.** Beaches with large tide ranges can be flat and wide, and –
as in the case of this one at mid-tide on Cable Beach, near Broome,
Western Australia – create perfect sunsets.

*Rob Brander*

**Pic. 16.** The first surging wave of the 2004 Boxing Day tsunami
crashes over the seawall at Ao Nang, Thailand.

*Wikimedia Commons user David Rydevik*

**Pic. 17.** Waves and surge from an East Coast Cyclone swamping Stanwell Park Beach in New South Wales.

*Rob Brander*

**Pic. 18.** A group of intrepid scientists (Jamie MacMahan leading the way, me fourth from left) and naïve volunteer students carry GPS drifters into the jaws of a rip. We had GPS under our caps to measure rip current escape strategies.

*Patrick Rynne*

**Pic. 19.** A release of harmless purple dye into a rip current shows that some rips can take you for a ride in a wide circle, eventually bringing you back into shallow water with the breaking waves.

*Rob Brander*

**Pic. 20.** On the other hand, this dye release at Sydney's Palm Beach shows that rip currents have a nasty habit of flowing well beyond the surf zone.

*Rob Brander*

**Pic. 21.** A classic channelised rip flowing along the shore and then out through the surf zone. White is nice, green is mean!

*Rob Brander*

**Pic. 22.** The dark channel of water next to the rocks is a boundary rip current at Fingal Head, New South Wales. Further up the beach are some channelised rips (dark gaps).

*Rob Brander*

**Pic. 23.** The arrow points to a flash rip on Sydney's Coogee Beach that has just developed, carrying turbulent water and sand offshore. There's a smaller one at the bottom of the picture.

*Rob Brander*

**Pic. 24.** Swash rips can form from the backwash from beach cusps on steep beaches like Pearl Beach, New South Wales.

*Michael Hughes*

Pic. 25. Sediment size has a huge control on the slope of the beach as shown in this picture of Cable Bay on New Zealand's South Island. The cobble section is steep, the sandy section is flat.

*Rob Brander*

Pic. 26. The sand on both beaches of Tomaree Headland in Port Stephens, New South Wales, is the same. The difference between them shows the influence of big waves versus small waves.

*Rob Brander*

**Pic. 27.** View from a hammock on my favourite beach. A steep and narrow reflective beach in Thailand. Plenty of time for reflection!

*Rob Brander*

**Pic. 28.** A wide and flat dissipative beach at Stanley, Tasmania. Whitewater everywhere, but no rips or sandbars.

*Rob Brander*

**Pic. 29.** A longshore bar and trough beach as evident by the prominent line of breaking waves across the sandbar running parallel to the beach like a highway.

*Rob Brander*

**Pic. 30.** Crescentic sandbars creating a rhythmic bar and beach at New Zealand's Tairua Beach on the Coromandel.

*Rob Brander*

**Pic. 31.** Surfers Paradise on Queensland's Gold Coast, Australia, showing transverse bars with channelised rip currents snuggled between them.

*Rob Brander*

**Pic. 32.** A long period of small waves will bring sand onshore, forming a low tide terrace beach characterised by small channelised rip currents, like the one in the middle.

*Rob Brander*

**Pic. 33.** Believe it or not, all this sand came from the beach on the far right. A massive coastal dunefield in Scott Bay, South Australia.

*Patrick Hesp*

**Pic. 34.** Pool closed for swimming! Severe coastal erosion at Sydney's Narrabeen-Collaroy Beach following an East Coast Cyclone in June 2016. The damaged section did not have a boulder seawall in place.

*Chris Drummond, UNSW Water Research Laboratory*

**Pic. 35.** A bluebottle jellyfish washed up on the beach. The bubble is about 10 centimetres in size and won't sting you. The thin blue lines are tentacles that can be as long as a metre and will sting you!

*Rob Brander*

**Pic. 36.** Stings from jellyfish can be painful, but the box jellyfish sting can be horrific and can scar for life. Pay attention to warning signs on tropical beaches where they occur and treat with vinegar if stung.

*Jamie Seymour*

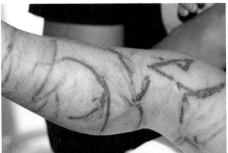

**Pic. 37.** Small, shy, cute and extremely deadly. The blue-ringed octopus is common in rock pool crevices and turns blue when unhappy. Do not touch!

*Shutterstock, photographer K.Pock Pics*

**Pic. 38.** Here's looking at you kid.

*Wikimedia Commons user Hermanus Backpackers*

# Boundary rips

When waves approach beaches at an angle, they can create a longshore drift of water along the shoreline that will flow happily along the beach until it hits a physical boundary where it is then deflected offshore. That's the easiest way to describe boundary rips, which are quite common on any beach that has breaking waves and natural boundaries such as headlands, rock platforms, rock outcrops or unnatural boundaries such as groynes, seawalls, jetties and piers. Boundary rips flow alongside these structures and because they are there almost all the time, they erode deep channels and also appear as darker gaps of water between the structure and the whitewater of adjacent shallower areas (Pic. 22). They've often been referred to as permanent rips, headland rips and topographic rips, but are also often given names by the locals (usually lifeguards and surfers) such as the 'Backpacker Express' at the southern end of Sydney's Bondi Beach. This rip is notorious for carrying young tourists out to sea where many of them get an unexpected cameo role in *Bondi Rescue*, a popular reality show about the lifeguards.

It's a good idea to avoid swimming next to any physical boundary on a beach because often you can get boundary rips forming on both sides of the structure. It depends on the wave direction. Let's say your coastline is oriented south to north and you've got two beaches separated by a headland. If the waves approach the coast from a southerly direction, they'll create a longshore drift that moves northward along the beach and creates a boundary rip on the southern side of the headland due to the water being deflected offshore. That makes sense.

## HOW TO SPOT A RIP | The best way to spot a rip current is from an aeroplane. While that advice is not going to be of any help to you, it has a message – it's much easier to spot rips from above. Looking down at a beach from a higher vantage point, like a headland, sand dune or even the carpark behind the beach will definitely help. Even the back of the beach is a bit higher. Having a pair of polarised sunglasses also helps 'see' through water. Being able to consistently and accurately spot a rip takes time and practice and can be difficult because there are different types of rips and rip flow can change quite quickly. But it's definitely possible to learn.

When you go to the beach, don't just run into the water; spend a few minutes checking out the conditions. Asking lifeguards or surfers to point out rips every time you go to the beach is also a great idea. Here are the most common visual clues that give rips away.

### 1. WHITE IS NICE, GREEN IS MEAN: CHECK FOR DARK GAPS | Most rips sit in deeper channels between sandbars. Not only is deeper water always darker, waves don't break as much over it. On the other hand, waves break in shallow water over sandbars. So look for narrow dark gaps, typically from 5 to 30 metres wide, that extend offshore between areas of breaking waves and whitewater (Pic. 21). Remember that 'white is nice' because whitewater means waves are breaking over shallow water and 'green is mean' because those green gaps are deeper and could be rip currents. Also remember that rips – the dark green gaps – can flow straight offshore, or at angles to the beach, and may even meander.

### 2. CHOPPY, DISTURBED WATER SURFACES | Rips move water offshore while waves bring water onshore so there is always a bit of interference between the two, creating a slightly bumpy, disturbed water on the surface of the rip. Not all dark gaps in the water are rips so look closely at the water surface texture. A rip will have a slightly rougher texture.

**3. CLOUDS OF SAND AND TURBULENT WATER |** Very strong rips, like flash rips, which tend to occur during large and messy wave conditions, can carry large amounts of suspended sand and flow past the line of breaking waves into deeper water. Look for pronounced rip heads full of sand and turbulent white, streaky water extending seaward off the breakers (Pic. 23). These usually don't last for more than a minute or two and can pop up all over the place.

**4. MOVING THINGS |** Rips move anything that floats, including seaweed, jellyfish, bubbles and people! If in doubt, throw a piece of driftwood or seaweed into the water and see where it goes. If you look carefully for a few minutes, you should be able to see the thrown object moving offshore.

**5. RIP EMBAYMENTS |** It's always a lot harder to spot rip currents from the shoreline, but it does help to look sideways along the beach, preferably from a bit of elevation. Often rip currents can remain in the same location for days or weeks and can carve out mini-embayments along the shoreline that are shaped like scallops. If you look along the beach and the shoreline is nice and straight and suddenly there's a pronounced mini-embayment, check to see if there's a narrow dark gap between whitewater heading out from it. If so, chances are it's a rip!

**6. TEACH YOURSELF! |** The more you go to the beach and ask people, like lifeguards or friends who are experienced surfers or beachgoers, how to spot a rip, the quicker you'll learn. Seeing rips in real-life is always the best method. However, there are a lot of pictures and movies of rips out there on the internet, although the quality and accuracy can vary. I hate to plug my own website, but if you go to <www.scienceofthesurf.com> not only are there links to videos of rip currents, but there's a Rip of the Month feature that I started way back in 2009. Every month I post a picture of a rip current and talk about it. That's a lot of rips! In fact, I wouldn't be surprised if it's the largest collection of rip current pictures in the world. Somebody needs to alert the *Guinness World Records*.

You might then think that it's safe to swim on the beach on the other side of the headland because it will be protected by the southerly waves. Good thinking, but this is where knowledge of how rip currents form can save your life. Yes, the beach at the northern side of the headland is protected by the southerly waves and there won't be many waves breaking there. However, the southerly waves will break on the middle of the beach and the water levels will be higher there than they are in the protected corner. Water will then start to flow south towards the headland where it's again deflected offshore. It's a double whammy! There are rips on both sides, and neither side of the headland is safe to swim. The same thing applies if the waves come from a northerly direction. So unless you are an experienced surfer looking for a free ride out the back, when it comes to swimming, stay away from any type of structure on the beach.

## Flash rips

It would be really nice if all rip currents occupied deeper channels and looked like dark gaps because that would make it easy teaching people how to spot them. Unfortunately they don't – thanks to flash rips. Flash rips are formed when a few large waves break and the water piles up locally generating a sudden, narrow offshore flow of water. This can occur close to the shoreline, but more commonly off the back of sandbars. Unlike channelised rip currents, flash rips vary a lot over space and time, popping up and disappearing here and there and never really lasting for more than a few minutes at a time. I remember meeting lifeguards on the east coast of the United States who called them 'popcorn' rips, which suits them perfectly.

Flash rips tend to be more common on days when there are bigger waves, or the waves are messier in general. They are dangerous because they are almost impossible to predict, form suddenly and can quickly take people offshore. When lots of people are standing on the seaward end of sandbars, flash rips can often result in mass rescues. It's even possible that a flash rip was the cause of the infamous drownings and mass rescue that occurred at Bondi Beach in 1938 (see page 97). They are also hard to spot because they appear as streaky, choppy, turbulent water often carrying clouds of sand and bubbles beyond the lines of breaking waves (Pic. 23). Sure you might be able to spot them from a headland or another vantage point, but that doesn't help you when you're in the water and one suddenly takes you offshore! Flash rips also won't circulate you back to the beach so it's best to think twice about swimming when conditions are messy and ask the lifeguards if there are any flash rips around.

## Megarips

Think big. Very big. Fast too. Megarips are the biggest and fastest rips of all, but they fortunately don't occur very often and only during major storms or swell when waves are huge. The massive amount of water brought in by these breaking waves can get funnelled into the middle of the beach or against a headland where it gets pushed offshore as megarips that can flow at average speeds of 2 metres per second for distances of more than half a kilometre. While this may sound terrifying, the good news is that you are extremely unlikely to be swimming during these conditions, so megarips are rarely dangerous to people. They do, however, cause major erosion to beaches by carrying sand a long way offshore.

## Swash rips

Swash is the combination of the uprush of water from a broken wave up the beach and its subsequent backwash down the beach. When large waves break on steep beaches the backwash can be extremely strong and carries water down the beach and out past the shorebreak, creating mini flash rips (Pic. 24). Swash rips don't go very far from the beach and don't sit in channels, but are enhanced in half-moon-shaped features called beach cusps, which are common along steep beaches. The problem about swash rips is that they can also move people, carrying them into deep water just beyond the shorebreak. For a non-swimmer, this could be potentially dangerous and this is probably what some people associate with 'undertow' as it seems like they were pulled out and under the breaking wave into deeper water.

## The bigger the rip, the bigger the trip

So just how far will rip currents take you away from the beach? Well as exciting as it sounds, they won't take you across the ocean from the east coast of Australia to New Zealand. It really depends on how big the waves are on the day. Rips will usually flow at least to the line of breaking waves and sometimes a little further offshore so the bigger the waves, the wider the surf zone, and the further offshore you'll go.

Under normal wave conditions, typical channelised rip currents along the south-east coast of Australia will take you about 50 to 100 metres offshore (Pic. 21). That doesn't sound too bad, but it all depends on how far you can swim in the ocean, which is a lot different than swimming in a pool! Boundary rips are a bit different and because they tend to be forced offshore, they can take you further offshore than channelised rips, even under the same wave conditions. Flash rips generally don't flow far, but because they often form off sandbars, you can still end up quite a way from the beach and in deeper water. However, some beaches always have massive surf and the rips are equally big. The rip that I floated out in at Muriwai Beach in New Zealand took me almost half a kilometre offshore and I was still going when the lifeguards pulled me out.

## Are rips friend or foe?

Rip currents are the main cause of drowning and rescues on surf beaches worldwide. While it's difficult to know exactly how many people drown, it's likely well over a thousand each year, given that rips occur on beaches in Australia, New Zealand, the United States, Central America, South America, Europe, South-East Asia ... well, pretty much everywhere you find sandy beaches and waves. In Australia, an average of 26 people drown in rip currents each year. That's more fatalities on average than those caused by bushfires, floods, cyclones and sharks combined!

This terrible drowning toll is perhaps not surprising because unless you live in tropical Australia, in which case you have to deal with crocodiles and deadly jellyfish, it's difficult avoiding rip currents on Australian beaches. Professor Andy Short from the University of Sydney's Coastal Studies Unit is the only person on the planet to have visited every one of the approximately 11 000 beaches in Australia and not only does he remember details about each one, he estimates that at any given time there are approximately 17 500 rip currents on them. Thankfully the most popular Australian beaches are patrolled by professional lifeguards and volunteer lifesavers and as long as you swim between the red and yellow beach safety flags put up by lifeguards, you'll most likely avoid rips. The problem is that there are plenty of popular unpatrolled beaches in Australia and unfortunately that's where most of the rip current drownings occur.

While drowning is clearly the worst-case scenario, just getting caught in a rip current can be a traumatic experience for both children and adults alike. Thousands of people are rescued from rip currents each year by lifeguards, lifesavers and surfers. Several years ago we conducted surveys and interviews with people who had been caught in rip currents and many of their experiences were harrowing and had turned them off swimming in the ocean ever again.

*There is always a slow … drift of water just above the seabed that is almost always moving offshore.*

On the other hand, rip currents are only dangerous if you get caught in one. So don't get in one in the first place! Always think about beach safety when you go to the beach. Learn how to spot rips and swim at beaches patrolled by lifeguards as much as possible. Ask the lifeguards about the surf conditions and always think about your own abilities relative to what's going on in the water. Do all these things and rip currents shouldn't pose any danger to you.

Rip currents also happen to be a surfer's best friend. It's a lot easier paddling back out to the waves in a rip than it is through the mess of breaking waves and whitewater. In some cases you don't even need to paddle! Experienced surfers and bodyboarders will always look for rips when they get to the beach because it's a free ride that can save you energy and extend your surfing session.

# Other currents

After reading all this information about rip currents, you probably won't want to step in the ocean again! Don't worry: not all beaches have rips and, even on beaches that do, they aren't there all the time. However, as long as waves are breaking and water is being brought to the beach, the water will still move around and there are two other types of currents close to the beach that you need to know about. They aren't nearly as much of a hazard to swimmers as rips are, but they are a very important control on how the sand on a beach moves around.

## Is there really no undertow?

If you've ever gone for a swim and been standing on a sandbar, you've probably had to backpaddle with your arms a little bit to stay in place as the water around your legs starts pulling you offshore. If you let yourself go with it, you'd slowly drift seaward, but not very far. It's not a rip, but it's definitely water moving offshore. There is always a slow, sometimes imperceptible, drift of water just above the seabed that is almost always moving offshore. It occurs pretty much everywhere across the beach and it's really no big deal to swimmers.

There's a bit of a debate among coastal scientists about what to call this phenomenon. Some call it 'undertow', which tends to mislead the public that they will be pulled under the water by something that doesn't actually exist. The more trendy and chic scientists are calling it 'bed return flow', which unfortunately is a term so hip that no one has a clue what they're talking about. So let's just call it offshore drift.

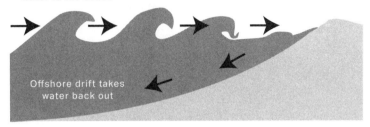

Breaking waves bring
water to the beach

Offshore drift takes
water back out

Not all the water brought in by breaking waves goes back out via rips.
Anywhere waves break, there will always be a gentle drift of water along
the bottom that heads out to sea.

When waves break across a surf zone, the whitewater from the broken waves heads to the beach in the top half of the water column and is balanced by offshore drift in the bottom half. It happens everywhere that waves are breaking, particularly around the shoreline and the crests and outer slopes of sandbars. If you look down at a beach from a headland and there are no rips, just lines of breaking waves and whitewater rolling towards the beach, you can often see clouds of sand slowly drifting seaward just behind the breaking waves. This is sand stirred up by the breaking waves being moved offshore by offshore drift.

## Longshore currents: always watch your towel

Have you ever gone for a swim on a very long beach and splashed and paddled around for half an hour only to find yourself further down the beach a long way away from your towel? Water can also flow along beaches, especially long ones, and this flow is called longshore drift. The drift normally

# ISLANDS MADE OF SAND

World Heritage-listed K'gari (formerly Fraser Island), in Queensland, Australia, is a prime tourist destination due to its pristine beaches, huge sand dunes, crystal-clear freshwater lakes bound by pure white sand beaches and a cheeky dingo population. At 123 kilometres long and 23 kilometres wide with dunes reaching an amazing 250 metres in elevation, it is also the largest sand island in the world. Further south, but equally impressive, is Moreton Island, which has Mount Tempest, the highest sandhill in the world at 280 metres.

What a shame all that sand doesn't belong to Queensland, but to its neighbouring state to the south, New South Wales. These islands are the final destinations for much of the sand moving up the east coast of Australia over many thousands of years driven by a northward longshore drift of sand caused by prevailing waves from the southeast. The sand only stops where it does because of some underlying geologic control and the fact that the Great Barrier Reef begins just a little to the north. The Reef blocks large waves reaching the coast, and effectively turns off the longshore currents, thereby bringing the drift of sand to a halt.

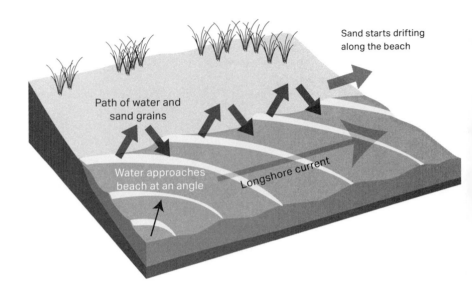

Sand starts drifting along the beach

Path of water and sand grains

Water approaches beach at an angle

Longshore current

Strong longshore currents that can move sand along a beach are formed when waves approach the shoreline at an angle.

refers to the movement of sand along a coast, which can create features like sand spits, barrier islands and huge sand islands like K'gari and Moreton Island in Queensland, but the sand can't move by itself. It has to be moved by longshore currents, which, just like rips, are created by breaking waves.

When waves approach the beach at an angle, the water from the breaking waves will also run up the beach at an angle and will either be pushed along the beach a short distance or will rush back down the beach in a straight line. If you start adding up all those little bits of water that have been pushed along, you get more and more water moving along the beach and before you know it you have a well-developed longshore current. Alternatively, strong winds blowing along the beach, such as the famous 'Fremantle Doctor', which kicks in during the summer along the Perth coastline in Western Australia, can literally push water along the beach, creating very strong longshore currents.

Longshore currents can be small, such as the feeders that can form part of rips, but on long beaches that stretch for several kilometres or more they can flow the entire length of the beach and can reach considerable speeds. Some currents flow as fast as rips, and while they won't take you out to sea, they can sweep you way down the beach if you aren't paying attention. For this reason, it's always important to check which direction the waves are coming from and to always look back at a fixed object on the beach, like your towel, a tree or a building, when you're in the water to make sure you aren't drifting away.

## Tidal currents

While the rise and fall of the tide can affect wave breaking and therefore the speed of rip and longshore currents, the tide itself does not usually create currents on beaches and if it does, its effects are almost imperceptible. Even king tides, the largest of the spring tides, will not generate significant tidal currents along a beach or in the surf.

Tidal currents are only really important on beaches close to tidal inlets near the mouths of rivers, estuaries or bays. As tidal inlets are generally narrow, the volume of water being squeezed through the inlets during the incoming (flood) and outgoing (ebb) tide can create strong currents, particularly within the inlet channels themselves. The greater the tide range, the stronger the currents will be.

While these currents are important for exchanging material between the estuarine environments and the ocean, they can also create tricky navigational conditions for boats and dangerous currents for swimmers. Tidal inlets are not the safest places to swim and not just because of the strong tidal currents. Apparently they are favoured locations of sharks!

# The bottom line

⭐ Water can move around in concentrated currents on beaches because of breaking waves. If there are no breaking waves, there are no currents.

⭐ Rip currents are the greatest danger to beach swimmers.

⭐ Rip currents are not undertow or rip tides, but are strong and narrow flows of water that extend from close to the shoreline offshore to the extent of breaking waves and sometimes beyond.

⭐ Rips flow faster around low tide and when breaking waves are bigger.

⭐ There are different types of rips and they have different appearances. It is important to know the types and learn how to spot them.

⭐ The most common type of rip is a channelised rip that sits in a deeper channel between shallow sandbars, making it appear as a dark gap through the surf. White is nice, green is mean!

⭐ Don't get sucked in by the rip: if you don't get in a rip, you won't drown in one.

⭐ There is always some water drifting gently offshore along the seabed.

⭐ Longshore currents flow along the beach and are caused by waves or strong winds hitting the beach at an angle.

⭐ Tidal currents are only dangerous to swimmers on beaches near river and bay mouths and in tidal inlets.

# 5

# Life's a beach ...

## But which one?

Is there such a thing as a perfect beach? I'd like to think so, but I'm not sure I've found one yet. Even if I did, I certainly wouldn't be telling anyone about it. Finding the perfect beach can be tricky, not because there are thousands of beaches in the world to choose from, but because beaches mean different things to different people. Ask a family with young children, a 21-year-old backpacker in search of a full moon party and a surfer to write down their top ten beaches and you'd get very different lists. The variety of factors that make beaches desirable is huge and it's really a matter of personal choice. The closest I've come to a perfect beach is a stunning little gem in Thailand called 'NAME WITHHELD AT AUTHOR'S REQUEST' where a lifetime spent swinging in a hammock between swims in the placid, warm tropical water would be pretty close to utopia for me. Bodysurfing on a perfect autumn day on Sydney's Tamarama Beach when the water is warm and the crowds are small comes pretty close as well.

Then again, I'm also partial to Dennis Port in Cape Cod, Massachusetts. When I was growing up, my family would drive from Toronto to the Cape every year for our annual holiday and we always stayed in the same motel on the same beach, eating at the same restaurants and meeting up with the same families doing the same thing. It was a tiny beach, backed by a seawall with rock groynes stretching for miles in both directions. The waves were small and the water was often cold and full of seaweed. Looking back, it wasn't much

*… a lifetime spent swinging in a hammock between swims … is pretty close to utopia for me.*

of a beach and some would call it ugly, but I loved it and still do. It's a sentimental thing and it will always be a perfect beach for me, if for no one else. My own list could go on and on, but the point is, there are perfect beaches everywhere and they are not really that hard to find, but they can be very different.

It's probably safe to say that for all those who love and use the beach in different ways, at some point a desire exists to understand a little bit more about what makes a beach look the way it does. The purpose of this chapter is to describe the common types of natural (and some artificial) features of beaches, explain why there are different types of beaches, what they look like and how they behave. Who knows, all this information may even help you find your perfect beach.

## The golden beach rules

If you want a simple, but meaningful definition of what a beach is, here goes: beaches are piles of sand (or sediments) that have been dumped there by waves. Sorry to disappoint, but that's it. The exciting part is really about how the sand moves around. Think about a beach that you visit every year, once a week or every day. Rarely does it look the same. It might be wider, flatter, steeper or narrower, or it might not even be there at all. If you really want to understand why your beach looks the way it does and how it changes over time, you need to know the three golden beach rules:

### Rule #1:
### *Waves stir up sand and currents move it away.*

Sand doesn't move by itself. It needs a little bit of help. If there's no wind or waves at all, the sand on the beach will just sit there doing nothing. However, once waves start to form, the gentle to and fro motion underneath them starts to exert a force on the seabed and at some point the first sand grains will start to roll and slide along. As waves increase in size and eventually begin to break, there is so much energy and turbulence that the sand grains are forced up into great big clouds where they remain suspended in the water for quite some time. If the sand clouds end up in a rip or longshore current, they'll eventually move away. In other words, as long as waves are breaking and currents are in motion, the beach is always changing because the sand is always moving.

### Rule #2:
### *Coarse sediments = narrow, steep beaches.*
### *Fine sediments = wide, flat beaches.*

Rocks don't float very well. Not even small ones. Throw a handful of pebbles in the water and see what happens. Not much! Being heavy, they'll sink quickly to the bottom. They don't travel very far and need a lot of energy to move them in the first place. For this reason, coarse sands, pebbles and cobbles tend to get slowly pushed up onto a beach, making it narrow and steep in the process. These sediments also have a natural tendency to sit at a steep angle. Therefore, you get a steep and narrow beach (Pic. 25).

The opposite is true of finer sand, which is good at floating around in suspension for short periods of time and can therefore be easily spread out by currents, thus making a wider beach (Pic. 25). Exceptionally powder-fine sand and silt can stay suspended for even longer and be carried further by waves and currents. This is one reason why fine sand and muddy beaches are often very wide and flat.

Did I say rocks don't float? Well, that's not entirely true. Pumice floats! Pumice is a type of igneous rock that is formed during explosive volcanic eruptions when hot magma cools rapidly. It's extremely light and full of holes, almost like a sponge, where gas bubbles were trapped, and this gives pumice a low density, which makes it float.

On most beaches on the east coast of Australia you can find pumice washed up near the back of the beach. Most people assume it's come from our volcanically inclined neighbour New Zealand, but most of it has actually come from an undersea volcanic ridge in the South Pacific near Tonga. Every now and then underwater eruptions release tonnes of pumice that makes its way to the surface, often as large rafts, where it's blown by wind and carried by ocean currents to the east coast of Australia. The pumice deposits don't really impact the beach, but pumice is a hard rock that makes an excellent skin exfoliant so bring one home for the shower!

*Rule #3:*
*Big waves take sand offshore*
*and small waves bring it back.*

In general, beaches with big waves tend to be wider than those with small waves. This is because big waves are good at moving a lot of sand around, particularly transporting finer sediments offshore, creating nice wide beaches. In contrast, smaller waves can't move as much sediment and mostly move it onshore, resulting in narrower beaches (Pic. 26).

This rule also applies to how beaches respond to storms. Big storm waves stir up a lot of sand and create currents that take it offshore, eroding the beach. However, storm waves don't last for long and the normal day-to-day smaller waves bring the sand back onshore. It might take months, but the beach will eventually recover. At least until the next storm hits and moves the sand back offshore again.

Remembering these simple rules will not only impress people at parties, but will help you explain a lot about why beaches look the way they do and how they behave. They also relate to the types of physical features found on your favourite beaches, starting with the most basic of all: their shape.

# The wonderful world of beach morphology

Contrary to popular belief, beaches are not just the dry bits of loose sand sitting above the shoreline that people like to lie on. Beaches actually begin some distance offshore where waves first start to feel the bottom and move sand around. This location can be a long way offshore when waves are big and closer to the shoreline when waves are small. Where do beaches end? On natural beaches, it's usually at the back of the beach where the sand dunes start. On urbanised beaches it's often a seawall, paved walking path or carpark.

If you were to cut a slice through a beach and look at it from the side, you would be looking at the beach profile as shown on the diagram on pages 120–121. Beach profiles all have one thing in common: they all tend to slope downwards in a concave fashion, being steepest at the upper part of the beach and becoming flatter the further they go offshore. Superimposed on top of beach profiles, you'll find all sorts of different forms, or beach morphology.

'Morphology' is a somewhat daunting Latin term commonly used by scientists, not to remind themselves that they are classically trained scholars, but because it sounds so much better than its literal translation of 'form science'. Beach morphology is thus the science of the different features on the beach.

Every beach on the planet has a beach profile and there are a lot of different processes going on across that profile. As a result, the beach profile is often split into sections that are given names, but unfortunately nobody seems to agree on the terminology. So I'm going to bend the rules of coastal geomorphology a little bit and split the beach profile into three simple parts: the nearshore, the surf zone and the beachface.

## The nearshore

The nearshore starts where waves start to slow down and change shape as they enter shallow water. There's not much to see out there other than small ripples on the seafloor, but it's a critical part of the beach because it often represents the

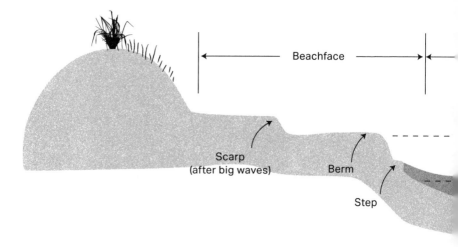

There's more to beaches than you may think
and they can be split into three sections, each
of which have their own different features.

dumping ground for sand taken offshore by currents during large storms. It's important because it represents a reservoir of sand that will eventually be returned to the beach by smaller waves.

## Surf zones and sandbars

The surf zone is where all the action takes place. It begins where waves start to break and ends at the shoreline. There is a *lot* happening in the surf zone as this is where most of the wave energy is released, stirring up sand and forming currents. The surf zone is also where you find sandbars and troughs (channels), commonly known as banks and gutters. Sandbars only really form on beaches that have significant wave action at least some of the time. Beaches protected from waves, such as those in bays and behind islands, often don't have a surf zone at all, simply because they don't have many

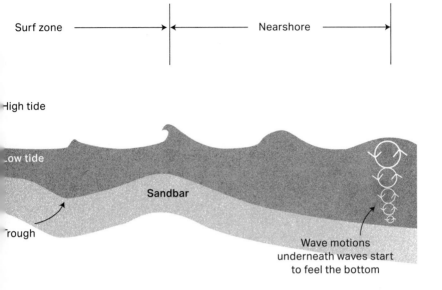

Surf zone ————————→ ←———— Nearshore ————————→

High tide

Low tide

Sandbar

Trough

Wave motions
underneath waves start
to feel the bottom

waves. Instead they just have a shorebreak where waves finally reach the beach.

Sandbars are piles of sand that can sit anywhere from tens to hundreds of metres offshore and are often separated from the beach by a deeper trough. They can extend along an entire length of beach or just exist on parts of the beach. They are sometimes difficult to spot, so the trick is to look for lighter areas of water, because the sandbar is in shallower water and reflects sunlight better than deeper water. Once again, wearing sunglasses with polarised lenses is an excellent way to 'see' the bars through the water. If there is also an area of breaking waves offshore and nowhere else, chances are it's a sandbar. Waves will break on sandbars more at low tide than high tide because of the shallower conditions. In this way, sandbars help protect the beach by taking the main impact of breaking waves.

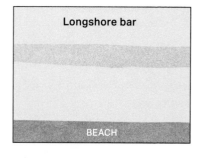

Longshore bar

BEACH

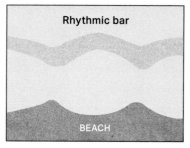

Rhythmic bar

BEACH

Transverse bars

BEACH

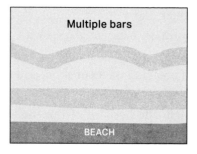

Multiple bars

BEACH

Sandbars come in
all shapes and sizes.
Viewed from above,
these are some of the
most common ones.

Dr. Rip's Essential Beach Book

Sandbars also tend to shift around a lot, which can quickly change the wave breaking patterns and surfing conditions. They come in a myriad of shapes and configurations and beaches can have one bar, several bars or no bars at all. Bars are described simply based on what they look like, as shown in the diagram opposite. Straight bars that run almost parallel to shore and are separated from the beach by a deep trough are called longshore bars. If the bar is slightly crescent shaped, it's known as a rhythmic bar. Sometimes the bars can be physically connected, or welded, to the beach and can stick out perpendicular to the beach, in which case it's called a transverse bar. Most surf beaches will have at least one offshore bar. If sand is plentiful, there may be multiple bars. You can also find multiple sandbars in low wave energy environments (with lots of sand) such as along many beaches in Port Phillip Bay, Victoria.

Despite how common they are, there's a bit of mystery about how sandbars form. The simplest theory suggests that because there is always a tendency for waves to slowly move sand onshore and for currents to move sand offshore, at some point the sand comes together and gets dumped in a big pile. As the waves and currents vary in size and strength, so too does the location of the bars, because all bars have a habit of moving, or migrating, both onshore and offshore.

Sandbar migration can be extremely rapid, as I experienced first-hand during my studies at the University of Toronto. I was fortunate to study with Professor Brian Greenwood, one of Canada's leading coastal geomorphologists, who every summer would take a team of keen students to help install equipment as part of his experiments on the behaviour of sandbars on beaches in the Great Lakes. Diving up to six hours a day in distinctly unpleasant 10°C water was hard, but Brian was a gifted motivator. He got us out of bed at the crack of dawn each day with inspirational cries of, 'It's flat calm out there, you buggers' and, 'It's never too late to change your marks, you know', and would proceed to bang and clang pots and pans at ear-splitting levels for a good half hour as he made his baked beans on toast.

During a storm, much of the equipment would often be buried under metres of sand by the moving sandbars. It was our job to excavate it with the equivalent of an underwater fire hose connected to an engine-driven pump onboard a small boat. This required wearing 35-pound weight belts with one diver blasting away with the hose, while another stood on his back to keep him down. We worked like moles, freeing cables and equipment in zero visibility, trying to ignore the fact that our regulators were jammed with sand, making simple things like breathing difficult. To make matters worse, the circa 1940s pump had a tendency to break down and the sand would quickly settle back in place, wedging us in like Chinese terracotta soldiers. We then waited until someone fixed it and we could blast our way out again. They were fun times and I learned a lot about the rapid speed of sandbar migration and the importance of occupational health and safety.

## The beachface: steps, berms and scarps

The beach that most of us know and love is the sandy part that sits above the shoreline. It is actually made up of two parts collectively known as the beachface. The upper bit is where most of us choose to set up our beach towels and is high and dry. The lower bit is the mostly wet intertidal section between the low and high tide marks and is dominated by the uprush and backwash of waves when they finally break at the shoreline. The combination of uprush and backwash is called swash, so the lower section is called the swash zone. The beachface also has some special morphology of its own.

Have you ever walked or run into the water only to suddenly trip, stumble and fall over? Just beyond the water's edge there is often a beach step, a sudden vertical drop that can range from just a few centimetres to more than a metre in height. It's not necessarily dangerous, but the drop can come as a surprise and can quickly put you in deep water, which might be a problem for non-swimmers or little children. Beach steps are caused by the turbulent interaction between the backwash rushing down the beach and the next incoming wave that is about to break.

At the top of the swash zone, near the high tide mark, there will often be a pronounced ridge of sand called a berm. Berms are formed when the uprush of a wave finally slows down, stops and dumps any sediment it may be carrying. Berms are important because they provide a line of defence *Berms are formed when the uprush of a wave finally slows down, stops and dumps any sediment ...* against storm waves and help protect the area of beach behind it from erosion. Although they are often completely destroyed by large waves, berms quickly build up again during lower wave conditions. If your beach has a well-developed berm, it means that you haven't had a storm for some time and there is a good supply of sand on the beach.

As it turns out, there is a relationship between how much sand is in the berm and how much is sitting offshore in sandbars. Many coastlines around the world experience some climatic seasonality and during a stormy season sand will get stripped from the beach and dumped offshore in sandbars.

During calmer months the bars will migrate shorewards and build up the berm and beach. So when you notice your beach seems to have a big berm and a lot of sand, there probably won't be many sandbars offshore and when your beach seems to have lost a lot of sand, don't worry, it's not gone forever, it's just sitting offshore in a sandbar.

On some beaches, the berm can evolve into a series of pretty half-moon shapes along the beach called cusps (Pic. 24). Cusps are almost entirely unimportant as there have never been any reports of a cusp migrating and burying a sunbather, but they look pretty and have proven both fascinating and frustrating to many coastal scientists because their spacing along a beach is uncannily regular. Trying to explain the almost perfect regular spacing of beach cusps has become a quest almost on the scale of the holy grail because there must be something neat going on to cause it. Other natural features, such as channelised rip currents, are also sometimes regularly spaced.

What we do know is that cusps seem to form when there are small waves and steep beaches. It's almost like a chain reaction. Once one cusp forms, a whole bunch form. We also know that one thing cusps, berms and beach steps all have in common is that the coarser the beach sediments are, the larger and more pronounced these features will be.

Finally, if you walk along the upper part of the beach and notice (or fall over) a vertical drop, almost like a mini-cliff, running along the beach, it's a sign that the beach has experienced some pretty significant and recent erosion. These notches, or beach scarps, are caused by large waves, very high tides, or a combination of both that eat into the beach, eroding the sand away. Most minor scarps are quickly re-worked on the beach by the action of waves and tides and the beach returns to normal, but during large storm wave events, the sand dunes at the back of the beach can be eroded, leaving large scarps behind. If scarping of the beach and dunes persists for a long time, it means the beach has a chronic erosion problem.

## Beach types

One of the best coastal scientists around, and a very good friend of mine, is Professor Gerd Masselink from the University of Plymouth in England. I first met Gerd during a backpacking trip to Australia where I helped him with his PhD research on the big-tide beaches of the Central Queensland coast. With lots of instruments poking out of the sand, we often had people coming up asking what we were doing. Gerd was happy to oblige, explaining that he was trying to measure the waves, currents and sediments to work out why these beaches looked different to other beaches. Most people seemed genuinely interested by this explanation except for one grumpy local, who listened carefully before walking away muttering, '*!*#*$! waste of time. Seen one beach and you've seen them all!' Gerd and I stared at each other dumbfounded, but if we'd had the wits to respond we would have said, 'Actually, that's not true …'.

Beaches may look the same, but take a closer look and you'll quickly start to notice subtle (and sometimes not so subtle) differences. For instance, the sand may be a different colour or size. One beach may be wider or steeper than another. Some beaches will always have big waves while others never do. The truth is that no two beaches are exactly the same. However, they can be similar. For this reason, you can find beaches in Australia, South Africa, Peru and California that look almost exactly alike, despite being vast distances apart. Something must be going on, right? As it turns out, there is.

Chapter 1 explained the importance of geology and sediments in creating sandy, gravel, muddy and carbonate beaches. Given similar geologic and wave conditions and the fact that all beaches adhere to the golden beach rules described earlier, it's not surprising that beaches can share many characteristics in common. Fortunately, we have a good understanding of what makes a beach tick and we can lump beaches into a number of broad categories, not only according to what sediment they are made of but what they look like.

## Steep and narrow beaches

Steep and narrow beaches can be found anywhere in the world, as long as they meet some of the following criteria. First, if a beach almost always has small waves, the sand will be piled up on the beach, making it steep (Pic. 27). Second, if there is not a lot of sand around, the beach will form and then pretty much stop growing, making it narrow. Third, if the sediments are made up of coarse material, be it sand, gravels or broken bits of coral, the beaches will also tend to be both steep and narrow.

These beaches are known as *reflective* beaches because some of the energy released by the breaking waves tends to bounce, or be reflected, back offshore by the steep beach profile. They are often recognised by the presence of beach cusps (Pic. 24), a pronounced beach step and the absence of sandbars and rips. This doesn't mean they are completely safe for swimmers and small children, because the steep profile promotes surging waves, which can knock people over, and plunging waves, which can cause a nasty shorebreak. The depth increases rapidly seaward of the shorebreak and there's not much happening, so you can have a nice swim. Definitely leave your surfboards at home though.

Reflective beaches are very common anywhere there are coarse sediments and/or protection from large waves. This includes beaches in bays, harbours and lakes as well as those behind offshore islands, and rock and coral reefs. They are found all over the world and also tend to be very stable because they don't change very much. Gravel beaches are always reflective and so are those idyllic 'coral sand' beaches on coral cays.

However, it's a mistake to assume that reflective beaches never get big waves. There are some pretty scary high-energy gravel beaches around the world that get pounded continuously by big waves. I once had a PhD student working on beaches like these in New Zealand. We felt he needed an immersive educational experience so when we surveyed the beach, off he went with the survey rod to stand in the shorebreak to get some measurements. We gave him a survival suit to wear, mostly out of pity, but this offered little protection from all the cobbles and pebbles that were flying around head height every time a wave broke. Fortunately he survived, but these beaches need to be respected. It's also important to remember that the sediments on gravel beaches are quite loose when covered by water and it's easy to sink into and lose your footing and often quite difficult to get back out. Swimming on these beaches is definitely tricky.

# Wide and flat beaches

Wide and flat beaches are almost the total opposite of reflective beaches because they have fine sands and/or big waves, or a combination of both. If a beach has fine sand it will automatically tend to be flat, and if the waves are continuously large the sand will be spread across the beach profile by waves and currents, creating a wide and shallow surf zone. These beaches are called dissipative beaches because waves start breaking way offshore and dissipate their energy over a wide area. Anytime you see a beach with a surf zone that is completely covered by whitewater, it's probably a dissipative beach (Pic. 28).

Some dissipative beaches have subtle longshore sandbars and deeper troughs, but rip currents in channels are usually absent. Instead, most of the water brought in by the breaking waves returns seaward everywhere by a gentle offshore drift. Just like reflective beaches, dissipative beaches are also very stable and tend to look the same pretty much every day of the year.

Dissipative beaches are found on any coastline that experiences consistently large waves, usually higher than 2 metres. For this reason, they are particularly common on beaches that are almost always exposed to high waves such as much of the coast of South Australia and the west coasts of New Zealand and South America. They are also found where major rivers reach the ocean because rivers are particularly good at delivering fine sediment to the coast. Brazil, for example, has a lot of big rivers, big waves and strong longshore currents that sweep the fine sediments up and down the coast. As a result, there are a lot of wide and flat beaches.

Are they safe to swim on? It's a tough call because they don't have many rip currents, but the large waves can make the surf zone extremely energetic, and paddling out for a swim or surf can be a heart-pounding, messy and ultimately unrewarding experience. Also, if the beach is fairly long there may be strong longshore currents in the troughs and you might end up with a long walk back to your towel.

It's also worth pointing out that not all dissipative beaches have big waves. There are plenty of wide and flat sandy and muddy beaches found in tidal flat and estuarine areas and other low-wave environments. Are they good for swimming? Not really. Are they dangerous? Not really, unless you have a deep-seated fear of crabs like I do.

## Beaches with sandbars and rips

For some reason, medium-sized waves and medium-sized sand are the perfect ingredients for creating a beach that has sandbars and rip currents. As these beaches fall somewhere between dissipative and reflective beach types, they are often blandly referred to as *intermediate* beaches. That's where the blandness ends because not only do these beaches look different, they are extremely dynamic, provide great surfing conditions and are the most dangerous to swimmers because of the rips. That doesn't mean you shouldn't swim at them. Almost all the surf beaches in Australia fall into this category and they are fantastic. You just need to be careful, know how to recognise rips and understand how they behave.

Intermediate beaches also change a lot and we give them more specific names based on what their sandbars look like. Fortunately there is sometimes a sequence behind how inter-mediate beaches change and evolve.

Let's say you have an intermediate beach with some sort of configuration of sandbars and rips. If that beach experiences a storm with big waves that lasts for a long time, a lot of the sand wrapped up in the beach and bars will get chewed up and rearranged. Really big storms will dump sand offshore, forming a sandbar running along the beach separated from the beach by a deep trough and may be broken in places by the odd rip current channel. Not surprisingly, this is called a *longshore bar and trough* beach (Pic. 29).

The beach may stay like this for a while, but if the waves get smaller, the sandbars will start to migrate towards the beach. As they do, they begin to curve, becoming crescent shaped as rip channels start to develop, forming a *rhythmic bar and beach* (Pic. 30).

If the waves keep getting smaller, the bars will continue moving towards the shore and the curves can physically start to merge, or weld, with the beach, creating a series of transverse bars separated by rip current channels that are now well and truly locked in place. This is called a *transverse bar and rip* beach (Pic. 31).

If the waves continue to get smaller, the sand will keep coming so that eventually even the rip channels start to fill up, forming what looks like a flat bench, or a *low-tide terrace* beach (Pic. 32).

*If the waves continue to get smaller, the sand will keep coming so ... even the rip channels start to fill up ...*

How long it takes to complete this cycle really depends on the variety of wave conditions and can take a few weeks or a few months, but it all comes down to how often there are major storms. At any point during the cycle, another large storm may occur and the whole cycle begins again. Since storms depend on the weather, which is hard to predict, it's even harder to predict what an intermediate beach will look like even just a week into the future.

There are other factors to consider as well, such as the orientation of a beach and its exposure to waves. One end of a beach may be protected by a headland and always have small waves and look like a reflective beach, whereas the opposite end of the beach may be exposed to large waves and look like a dissipative beach. In the middle, the beach may have lots of bars and rips. For this reason, long beaches can have several different beach types at the same time.

## Long beaches and pocket beaches

Speaking of long beaches, why do some beaches stretch for 10 kilometres, while some are only a few hundred metres long? According to Professor Andy Short, the best source of information on Australian beaches, the average length of the 10 685 beaches in Australia is 1.37 kilometres. Interesting. I think a lot of people, especially tourists, think that Australia is one big, long sandy beach, but it's not. Once again, the major factor determining how long beaches are is geologic control. For really long beaches to form there needs to be a long gap between rocky headlands and there also needs to be a lot of sand around. In the case of Australia, almost half of the coast can be considered to be a rocky coast with plenty of bedrock outcrops and headlands giving us relatively short beaches. A lot of headlands are spaced very close together and the beaches occupying the embayments between them are small and referred to as pocket beaches. Depending on the beach sediments and the wave climate, both long and pocket beaches can be any of the types previously described.

# Big-tide beaches

If you travel up the east coast of Australia, you will find great surf beaches all the way north to K'gari. After this, you might as well chuck your surfboard, because the beaches become wide and flat with piddly little waves. The reason for this is that the Great Barrier Reef pops up offshore and effectively blocks big ocean swells from reaching the coast. It also squeezes and amplifies the tidal wave, creating some big tides and, as it turns out, the tide has a huge impact on what beaches look like (Pic. 15).

For example, think about an intermediate sandy beach with a fairly small tide range of less than 2 metres. This means that the location of the surf zone will only shift back and forth a little bit during the tidal cycle. This gives the waves and currents plenty of time to exert their magic on the same patch of beach, creating bars, rips and other features. Now imagine that the same beach has a large tide range of 4 metres or more. The surf zone would now be shifted backwards and forwards across a much wider distance very quickly and the impact of the waves and currents would be spread all over the place. This can bulldoze a beach, leaving a wide, flat and fairly featureless beach at low tide. Or it might create some subtle bars and channels that look like ponds at low tide and slowly drain water from the beach as the tide ebbs. These are called ridges and runnels.

Most of the big-tide beaches in Australia are in the northern half of the continent and are not suitable for swimming as they can be quite muddy and prime habitats for crocodiles and deadly jellyfish. However, there are plenty of big-tide beaches around the world where you can swim, albeit with big waves and rip currents. The northern coast of Cornwall in England

# THE LONGEST BEACH IN THE WORLD?

There are a lot of people out there who like to go for nice, long walks on the beach for exercise and just to experience the beauty of the ocean. Of course, the longer the beach, the better the workout and connection with nature. So just how long do beaches get if you really want to reach nirvana? Believe it or not, there are only three beaches in Australia longer than 100 kilometres: Ninety Mile Beach in Victoria and Eighty Mile Beach in Western Australia are both about 222 kilometres long, while The Coorong, also known as Ocean Beach, in South Australia is 212 kilometres. In New Zealand the longest beach is Ninety Mile Beach in Northland. But all of these beaches are dwarfed by Praia de Cassino in southern Brazil, which stretches for 254 kilometres. That's a long walk. Probably best to wear a hat and bring some water.

Who checks all this stuff out? Good question. If you do some metric conversions there are a lot of beaches with distances of dubious accuracy in their names. Might be time for some beach re-naming?

has tides of up to 8 metres, but also receives waves of up to 3 metres in height, creating some impressive, if chilly, surfing conditions. The south-west Atlantic coast in France is also a fantastic surf coast despite having tides of over 4 metres. Both of these coasts have pronounced channelised rip currents that are often fully exposed at low tide and fully submerged at high tide making them relatively harmless, but around mid-tide, water levels and wave breaking patterns fire them up for a short time, often resulting in mass rescues of unsuspecting bathers on busy days!

## Sand dunes and beaches

Many beaches are backed by beautiful sand dunes and just as there is a transfer of sand between the beach and sandbars, there can be an exchange of sand between the beach and dunes. There is a scene in the classic 1966 surfing movie *Endless Summer* showing surfers running for ages over sand dunes until they reach a final crest that reveals a fantastic undisturbed stretch of coast and a perfect surfing break at Cape Saint Francis in South Africa. The 1994 sequel, *Endless Summer II*, recreated the same scene, but when the surfers reached the same lookout, they were greeted by a development of houses and the surf break was a pale imitation of its former self. The reason? The houses were built over the dune sand that used to blow into the ocean during offshore winds and helped to create the once-perfect surf break.

If beaches are piles of sand formed by waves, coastal dunes are piles of sand formed by sand blown off beaches by wind. They are extremely important and sensitive environments. Dunes provide the final buffer of protection from storm waves. If you remove, or build on, a dune in order to get a good view of the ocean, you'll soon be in deep trouble as you may get more of a water view than you bargained for.

Do all beaches have dunes? Nope. For dunes to form there needs to be a lot of sand around, plus a lot of wind, and the sand has to be fine enough to be able to be blown by the wind.

Most important to coastal dune formation is the role of vegetation, which acts to trap sand and bind it in place, thereby building the dune and stabilising it. Dunes can grow quickly, starting with a low incipient foredune at the back of the beach that is vegetated by dune grasses, such as marram or spinifex. Incipient foredunes are very fragile and can be wiped out by a single storm or even by just a few vehicles driving over them. If they become colonised by small shrubs and trees, however, they can grow in size and become much larger, semi-permanent foredunes, which may grow up to more than 10 metres in height.

All coastal dunes rely on vegetation to keep them stable, and whether it's erosion by storms, rips or human impacts, weaknesses can occur, causing gaps in the foredune called blowouts. Sometimes the blowouts can start moving inland, becoming parabolic in shape, and can start to bury vegetation and roads and can even threaten houses. For this reason, it really is important to pay attention to those 'Please Stay off the Dunes' signs. The vegetation is there for a good reason.

There is also a strong relationship between the type of beach and the size and type of sand dune behind them. Reflective beaches generally have a small well-established foredune because of their low waves and stable nature. Beaches with bars and rips often have blowouts because of the impact of larger storm waves and rip currents. Dissipative beaches, with their high wave exposure and wide beach, are often backed by massive dune fields that can extend for kilometres inland

## AN IDIOT'S GUIDE TO DRIVING ON A BEACH |

I'm sorry if the title offends anyone, but when it comes to driving on a beach, I'm an idiot. One of my mentors, Professor Andy Short, has been driving on beaches all his life and claims to have never been bogged, but in an amazing twist of irony, I've been bogged on almost every beach I've ever driven on. I'm sure Andy will have a choking fit when he reads this, but I've learned quite a lot about driving on a beach (mostly from him). Don't use a Toyota Corolla hatchback for a start. Beach driving is the realm of four-wheel drive (4WD) vehicles. Even so, it's easy to get stuck, so here are some tips.

**BEFORE YOU HIT THE SAND**

★ Get local advice on beach conditions, check the tide tables and only use established access routes to the beach.

★ Reduce the tyre pressure for soft sand driving to give better traction and reduce track erosion. Letting them down by 15 psi is a good rule of thumb.

★ Make sure you have a pressure gauge and a device for re-inflating tyres when you return to a hard surface or else you can damage them.

**ON THE BEACH**

★ Normal road rules apply.

★ Don't speed. Hazards such as deep pools, drainage gullies and rock outcrops are hard to see.

★ Drive smoothly in low range with gear changes at high revs.

★ Avoid sharp turns, rapid accelerations and sudden braking. Coast to a stop and make turns as wide as possible.

- ★ Try to drive within two hours either side of low tide and between the waterline and high tide mark. Sand is harder and firmer here.

- ★ Don't drive in salt water, to avoid soft sand and salt damage.

- ★ Don't blaze your own trail. Use existing beach tracks or follow the vehicle in front as the sand will be compressed and harder. NEVER drive on vegetation.

- ★ When driving over dunes, drive straight up or down and give way to downhill traffic on tracks.

- ★ Park at an angle to the shoreline, on a downhill slope facing the water. This helps to start and move away easier. Always start with wheels pointing straight.

- ★ Keep your speed. In other words, avoid losing your momentum and avoid soft, loose sand wherever possible.

## IF YOU GET BOGGED

- ★ DON'T floor the accelerator, you'll only dig yourself in deeper.

- ★ Remove the build-up of sand from behind the tyres and underneath the vehicle.

- ★ Try to reverse in your own tracks. This may involve going backwards and forwards carefully ('rocking') to compact the sand and avoid digging in.

- ★ If this isn't working, let down your tyres by another 12 psi and have another go. After this, drop it by smaller increments, but try not to go below 10 psi.

- ★ People pushing really does help.

- ★ If it's bad, find someone who knows how to use a snatch strap, winch or jack! My friend Dave Mitchell says to bury a tyre and winch yourself out using that!

and can look just like desert dunes. Some of the best examples in Australia are along the high-energy coast in southern Australia (Pic. 33) and a whole series along the coast of New South Wales from just north of Newcastle to Seal Rocks. For an international flavour, my pick is the Grande Dune du Pilat in Bordeaux, France, which is the largest sand dune in Europe.

# Human impacts on beaches

## Cars, groynes and seawalls

There is something inherently wrong about driving on the beach. This thought occurred to me as I careened wildly out of control in my Toyota Corolla hatchback on Himatangi Beach on New Zealand's North Island. I was following my good friend and coastal dune expert Patrick Hesp, who wanted to show me some features of the beach and claimed that 'I'd be fine', advising me to 'keep the revs up'. This was probably good advice had I been driving a four-wheel drive like he was, rather than a compact car. Within seconds of hitting the beach, pieces of driftwood, some still attached to washed-up tree trunks, flew up and over my windshield before wrenching themselves into the undercarriage, making hideous grinding noises. Trying to 'keep my revs up', my brain abandoned all previous knowledge of how to change gears, and I avoided stalling only by flooring it in first gear. I motivated myself by visualising people frantically trying to dig out a car while waves swamped them – the sort of situation you never wanted to find yourself in. Unfortunately, I was in danger of becoming one of those people.

After spraying large walls of sand in all directions and carving deep ruts in the beach, I managed to turn the screaming and bucking car around and made it back to safety, much to the amazement of shocked onlookers choking in the smoke belching from my car. At least I could take some solace from the fact that the next day, all the damage I did to the beach would be erased by the high tide. In truth, unless you are driving over fragile dune vegetation and sunbathers, driving on the beach is not really harmful to the beach itself. It can get very ugly though and it can be disappointing to arrive at a beautiful beach to discover it's a carpark full of deep-rutted tyre tracks.

Building on a beach is a whole different ballgame. As soon as you build any sort of structure on a beach, on a dune or in the surf zone, everything is affected by it: the waves, the patterns of sand movement and the shape of the beach itself. Sometimes these changes are insignificant, at other times they can lead to a chain reaction, or 'domino effect' of problems. There is an old adage among coastal scientists that says, 'there is no erosion problem on a beach until a structure is built'. It can be very true. Beach erosion is a completely natural process. No one seems to jump up

*... pieces of driftwood, some still attached to washed-up tree trunks, flew up over my windshield ...*

and down and complain when a pristine beach in a natural park is badly eroded by a storm, because generally it will fix itself. However, when some luxury homes and resorts are built too close to the beach, all of a sudden there is mass hysteria about the 'erosion problem' and cries for action to fix it.

The truth is, it's almost impossible to fix beach erosion. Many coastal erosion problems along popular stretches of beaches can be traced back to the early 1900s when beach living and tourism started to gain popularity and a flood of coastal development began. The damage to valuable property by storm waves and rising tides was almost immediate. However, these erosion problems were often related to either mismanagement of natural coastal dune systems, building too close to the beach in the first place, or building in totally inappropriate coastal locations.

So what can be done? One approach is to do nothing and let nature take its course. If a few houses fall into the ocean, it's not the end of the world. However, this is not a wildly popular approach for the homeowners in question and is often unrealistic. Another approach is to retreat and relocate. The idea here is to either physically remove valuable or historic structures, such as houses or lighthouses, further back from the ocean. Some local governments actively attempt to buy back coastal property in order to get rid of the structures entirely and re-establish the natural coastal dunes. This is a difficult, expensive and very slow process because not everyone wants to sell their beachfront property!

For these reasons, the traditional response to protecting property from an eroding beach has been to build other structures to try and outsmart and hold back the advancing waves. This hasn't been as easy or as successful as hoped and on some beaches has made the situation worse. Groynes are structures, usually made out of piles of big rocks, that stick out into the ocean perpendicular to the beach in order to trap sand moving in one direction. That's great for one side of the groyne as a nice wide beach forms. The only problem is that the downdrift side is cut off from the sand and loses its beach. So what happens? Another groyne is put in and then another … and another. Not only do the groynes become useless and redundant after a while, but the beach starts to look awful. If you have a beach left at all, that is.

Seawalls are fairly vertical structures typically built out of concrete, boulders, steel or wood that are designed to stabilise the shoreline and protect the area or property behind. They are usually very good at this. What they are not good at is fixing a beach that is eroding. The reason you don't find big healthy beaches in front of natural headlands, rock platforms and rock outcrops is because there is so much wave reflection and turbulence in these areas (Pic. 14) that the sand can't stay there and gets washed away. Seawalls can have a similar effect, so building a seawall might save your property, but it won't be doing the beach any favours.

Not only can the direct effects of groynes and seawalls be dramatic, but there can also be adverse side effects considerable distances away from these structures. They have a habit of breaking down over time under the impact of waves, leading to the construction of even bigger seawalls. After a while, your beach can start to resemble a permanent construction zone. Believe it or not, the term used to describe this process is called the 'New Jerseyisation' of a beach.

New Jersey is a state on the east coast of the United States and has some of the most popular beaches in the country. Tourists love it, but an argument can be made that in some places they've loved it to death. A small town named Cape May became America's first real beach holiday destination in the early 1800s and as word spread, so too did an initial wave of unbridled development along the New Jersey coast. Everyone was happy until storm erosion started to occur, causing damage to buildings and roads. Seawalls and groynes were built to address the erosion, but meanwhile more people kept coming and development increased even more to account for all the people. The storms kept coming too, further eroding the beaches, damaging the groynes and seawalls and destroying houses.

Over the years, the answer was to build bigger groynes, bigger seawalls and taller hotel blocks behind them to see over all the construction at a beach that no longer existed. The old beach was really just a pile of rubble. This whole process is where the 'New Jerseyisation' of a beach comes from, but examples of inappropriate coastal development and management practices exist around the world, including Australia. I should mention that I once had a student from New Jersey who kept sending me pictures of beautiful beaches from there when she went home, so it's not all bad news!

Before you start jumping on the anti–coastal engineering bandwagon, it should be remembered that not all groynes and seawalls have been disasters and in many cases have been the only viable option available. This was evident along Sydney's Narrabeen-Collaroy Beach where chronic beach erosion has occurred for years. In a severe East Coast Cyclone in June 2016, the beach eroded back by more than 50 metres in places, damaging houses and causing an in-ground pool to end up on the beach (Pic. 34). What was interesting was that the entire beach eroded, exposing an old seawall made of boulders. The only stretch that didn't have the boulder seawall was where all the property damage occurred. So the seawall did its job – it protected the property behind it.

The reality is that erosion of beach property and coastal structures simply can't be ignored. Nevertheless, seawalls in particular remain a controversial issue as they generally designed to protect the assets of a select few, often at the expense of the beach and the thousands of people who used to enjoy it. Working with eroding beaches will always be a major challenge. Thankfully, beaches are remarkably resilient and are very good at healing themselves given the chance. There are also softer approaches to managing the coast that are more natural and aesthetically pleasing. The first is to rehabilitate and re-establish a solid line of sand dunes in front of threatened property. The other is to put sand back on the beach.

## Artificial and nourished beaches

Some of the most famous beaches in the world – including Miami Beach in Florida, Waikīkī in Hawai'i, the Costa del Sol in Spain, Copacabana Beach in Rio de Janeiro and, okay it's not that famous, but let's include it anyway, Brighton Beach in Melbourne – all have one important thing in common. They're all fake. Well, not exactly, it's just that much of the sand on them wasn't there to begin with and came from somewhere else.

Beach nourishment is a coastal management practice involving the addition of sand to a beach that is experiencing severe coastal erosion in order to widen it, or to create an entirely new beach that wasn't there before. In the case of extravagant coastal developments like 'The World' in Dubai, which is an island complex made to look like a map of the world, the beaches are completely artificial. However, in most cases nourishment takes place to address an erosion problem and, aesthetically, nourishment is a much better option to combat erosion than building groynes or seawalls. But it is not without its challenges, starting with the fact that it's very expensive.

As an example, Miami Beach had virtually eroded away in the 1970s, but was nourished at a cost of what today would be valued at over US$1 billion. Nourishment is costly because the sand has to come from somewhere and involves dredging, pumping and large-scale construction on the beach. It's a long and messy process and Miami's nourishment took five years, but at least there's now a long, wide beach there. The problem with many nourished beaches, however, is that they are prone to rapid erosion during storms and often need sand top-ups at additional expense. So it's not a quick and easy one-off solution. Such is the case at Melbourne's Brighton Beach, which may not have the mystique of Miami Beach, but is still a popular beach destination with famous and colourful bathing boxes, and has been undergoing nourishment for over 30 years due to ongoing erosion. Still, despite the effort and cost, at least at the end of the day, nourishment provides you with a beach to lay on, and that's what makes most people happy.

# The bottom line

★ Beaches are always changing because waves stir up sand and currents move it away.

★ The coarser the sediment on a beach, the steeper and narrower the beach will be; the finer the sand, the wider and flatter the beach will be.

★ Big waves and storms take sand away from the beach; small waves bring it back.

★ Sandbars come in a range of shapes and sizes and can move both onshore and offshore.

★ The amount of sand on the beach is related to how much is offshore, usually in sandbars.

★ The coarser the sediments, the more pronounced features such as berms, steps and cusps will be.

★ Beaches with medium-sized sand and medium-sized waves have sandbars and rips.

★ Protecting sand dunes and nourishing beaches are the best ways to manage beach erosion.

★ Building seawalls and groynes is usually bad news for beaches.

# 6

# Shark fin soup and other things that can kill you

## Beach hazards and safety

The late Peter Benchley has a lot to answer for. The author of the best-selling novel *Jaws* was a staunch supporter of the conservation of sharks and later did his best to dispel their image of being 'man eaters', which is fantastic except for the fact that his book and the subsequent 1975 movie of the same name effectively traumatised a generation of swimmers, preventing them from ever stepping foot in the ocean.

I was ten years old when the movie came out and my parents did a wonderful thing. They refused to let me see it, and to this day I don't think about sharks when swimming in the ocean. This is in direct contrast to my Canadian friend Pierre who did see the movie. When Pierre visited me in Sydney years ago, I took him boogie boarding and found it odd that he only ventured as far as the shorebreak and was continually getting pummelled headfirst into the sand. When I told him to paddle out further to where the best waves were he said, 'Are you crazy? There's sharks out there!'

The tragic thing is that Pierre is not alone in his paranoia. Millions, possibly billions of people, are terrified of sharks. Every shark 'attack' becomes headline news. After a while you start to get the impression that sharks exist only to terrorise and eat people. Whenever I'm asked to give a beach safety talk to primary school kids, I always ask them how many people they think are killed each year by sharks in Australia. Usually there's one student who puts up their hand and says 'Ten thousand?' When I tell them that on average only about two people are killed by sharks each year, the response is stunned disbelief. It's as if someone told them there was no Santa Claus.

However, a school student once raised his hand and told me that more people are killed every year while eating shark than are actually eaten by sharks. I looked it up and couldn't find any information, but sure, why not, it would not surprise me if people have accidentally choked to death while eating shark fin soup. Anything is possible and it makes for a catchy chapter title, but the likelihood of it happening, just like getting bitten by a shark, is so incredibly small that it's really not worth worrying about. Instead, this chapter focuses on the more common and realistic hazards that you are more likely to experience when you go to the beach and what you can do to make sure your visit is a safe one.

*... more people are killed every year while eating shark than are actually eaten by sharks.*

# Staying safe the easy way

The main things to think about when planning to visit a beach are far less sinister than sharks, but much more important. When you first step foot on the sand, you need to ask yourself two questions:

1 What have I done to protect my skin against the sun?

2 Is it safe to swim?

Making wrong decisions about these simple questions can have very serious consequences, but making the right ones are easy and can lengthen your lifespan considerably!

## 'Slip, Slop, Slap, Seek and Slide'

Sorry folks, the carefree days of lathering up in suntan lotion for a day at the beach to work on your tan are over. You might think a tan looks good on you when you're young and trying to impress, but the more you sunbake, the more likely your skin will look like a wrinkled and saggy leather handbag by the time you're 40, which is definitely *not* a good look. The only thing you are really likely to gain from prolonged sunbathing is premature ageing and skin cancer.

Skin cancer is the darker side of tanning and is the most common type of cancer in many countries including Australia and New Zealand. Two out of three Australians will develop a form of skin cancer and approximately 2000 die each year from it and, as usual, it's men who are most at risk. Unfortunately, the numbers are increasing. Some skin cancers, such as melanoma, can go unnoticed but are particularly aggressive and spread easily. If you have spent a lot of time at the beach in your teens, make sure you get regular skin checks done when you're older. The sooner the better.

Skin cancer is caused by exposure to ultraviolet radiation (UV) from the sun, so the best way to protect yourself is to cover up as much as possible, particularly during the middle of the day when the sun is strongest. Don't be fooled by cloudy and hazy days – you can still get burnt! There's a great slogan from the Australian Cancer Council called 'Slip, Slop, Slap, Seek and Slide', which covers all the bases when it comes to sun protection:

» **Slip** on a shirt (long-sleeved shirts and rash vests are best, especially for kids)
» **Slop** on some sunscreen (at least SPF 30+ and reapply regularly)
» **Slap** on a hat (wide brimmed is better than a cap)
» **Seek** some shade. These days beach shelters are all the rage!
» **Slide** on a pair of sunglasses (preferably UV resistant and polarised).

It didn't take me long being in Australia to get the 'Slip, Slop, Slap' message, but 'Seek' and 'Slide' are new additions that make perfect sense. Stick to these rules and they'll soon become second nature and not only are they good for your long-term health, you'll be able to go home from a day at the beach without feeling like you've been roasted to a crisp. It's also a good idea to drink lots of water to avoid having a pounding headache due to dehydration. I do worry about the next generation though and wonder if they've heard 'Slip, Slop, Slap' so much that perhaps they've lost interest. It always drives me crazy when I take my students on field trips to beaches. I tell them numerous times before the trip to bring a hat, put sunscreen on and drink lots of water. Then in the blazing hot sun on the beach there they are – no hats, begging me for sunscreen, asking for water and looking like tomatoes in the evening. It's a painful lesson for them. The sun is no joke.

## Lifeguards and beach flags

If it weren't for lifeguards, the number of drownings on the world's beaches would be horrendous. The absolute safest place to swim on a beach is near a lifeguard. While lifeguards can't be everywhere all the time, fortunately the most popular beaches are the ones most likely to have lifeguards, and it really pays to swim at these beaches.

In most countries the presence of lifeguards is obvious by the presence of lifeguard towers, vehicles, rescue equipment and beach flags. Beach safety flag systems vary around the world. For example, Australia, New Zealand, the United Kingdom and South Africa use a pair of red and yellow flags on the beach to identify safer, supervised swimming and bathing areas. The core safety message in these countries is to always 'Swim Between the Red and Yellow Flags'. The flags are placed as far away from rip currents and other obvious surf hazards as possible and are under surveillance by professional lifeguards and, in the case of Australia, volunteer lifesavers who are there on weekends and holidays during the extended summer period. Some beaches have both lifeguards and lifesavers on duty on summer weekends so there are good reasons to always find the flags and swim between them!

Other countries, such as the United States, use a coloured flag system that indicates the relative safety level of the surf. For example, in Florida and other states, a red, yellow or green flag means high, medium and low surf hazard levels, respectively. These levels are also related to rip current warnings issued by the National Weather Service based on a similar hazard ranking. A double red flag means the beach is closed to swimming due to hazardous conditions. Hawai'i uses a series of flags that have different colours and symbols for beach and surf conditions with a yellow square for caution, a red stop sign for high hazard and a black diamond for extreme hazard. Years ago while visiting Vietnam, I went swimming at a gorgeous long sandy beach near Hoi An and spotted a single lifeguard station that consisted of stacked up old car tyres with a plastic chair on top flying the skull and crossbones! Whatever works, I guess. However, as lifeguard flag systems can vary from country to country, and even within countries, it's important to find out what they mean, and if you're not sure talk to the lifeguards.

## Unpatrolled beaches and safety signs

Who's keeping you safe when you go to the beach for a swim? The lifeguards are, so there's nothing to worry about … right? Wrong. Unfortunately, there are a lot of popular beaches out there that are completely unpatrolled. For example, Australia has almost 11 000 beaches, but less than 5 per cent are patrolled by professional lifeguards or volunteer surf lifesavers and the vast majority of these are only patrolled seasonally during the warmer swimming months. The same is true for beaches in the United States, with the exception of some in Florida and southern California that are patrolled year-round. It's therefore not surprising that most beach drownings occur on unpatrolled beaches, on sections of beaches a long way from lifeguards or outside of patrol hours.

**THE THINK LINE |** The main message I have been communicating to people during my public Science of the Surf talks has always been how important it is to go the beach and spend a few minutes thinking about beach safety. My mantra has always been that you don't cross the road without looking both ways and you should never visit any beach, whether it's patrolled by lifeguards or not, without thinking about beach safety. Those few minutes could save your life.

Every summer in Australia there is a terrible drowning toll on our beaches, mostly involving people caught in rip currents on unpatrolled beaches. The message that is repeated over and over again by authorities is for people to please swim only at patrolled beaches between the red and yellow flags. It would be fantastic if everyone listened to that message, because it's extremely rare for someone to drown between the red and yellow flags, but unfortunately I think the message is falling on deaf ears. Studies have shown that even though well over 90% of Australian beachgoers know that they should swim between the flags, many choose not to. So we have to do more. We can't just keep relying on the swim between the flags message. The reality is that people will always visit unpatrolled beaches.

Surf Life Saving Australia is the leading beach safety organisation in Australia and recently came out with a national safety campaign called the 'Think Line' that I believe is the next big step in reducing the drowning toll on our beaches. The 'Think Line' was developed to focus on the rip current hazard with the idea being that everyone should STOP and draw an imaginary line in the sand when they go to the beach and LOOK for rip currents and have a PLAN if somebody gets in trouble.

The 'Think Line' is a fantastic message that not only applies to rip currents, but all aspects of beach safety and it should become ingrained in our beachgoing culture so that we do it automatically whenever we go to the beach. STOP and think about whether it's safe to swim. LOOK for any hazards or lifeguards. Make sure you have a PLAN if something goes wrong. The 'Think Line' can also be applied to any beach in the world. So there's your answer – always think about beach safety and instead of making it an imaginary line, why not get creative and get the kids to draw a nice artistic line in the sand. That'll help get the message across even more!

STOP LOOK PLAN

There are many reasons why people visit beaches without lifeguards. Often it's just a matter of convenience because it's the beach closest to your holiday accommodation or where you live. A lot of people also like seeking out beaches away from the crowds and the fact that social media loves to promote things like 'Top 10 Secret Beaches' doesn't help either because they're not secret any more. There's a beautiful beach in northern New South Wales called Dreamtime Beach that used to be a quiet spot frequented almost exclusively by locals. It then became a social media darling and was consistently ranked as one of Australia's best beaches. This brought in hordes seeking the perfect selfie who also decided to go for a swim. As it turns out, there's a nasty boundary rip at the northern end and five people tragically drowned over a three-year period. Dreamtime Beach is completely unpatrolled.

So what is keeping you safe when you decide you want to go swimming at a beautiful unpatrolled beach? Most local governments put up various types of safety signs at public access points. These signs tend to come in all different shapes and sizes with all sorts of different warning messages about beach hazards like rip currents, but numerous studies have shown that safety signage is not as effective as authorities would like to believe and most people just ignore them. Ultimately the main thing that is keeping you, or your family, safe at the beach is YOU. Your own knowledge of beach and surf hazards and your ability to recognise them is extremely important when it comes to beach safety. So too is your awareness of your own abilities and the decisions you make about whether it's safe to swim or not. As the saying goes 'If in doubt, don't go out'.

## Watch out for fin chop

Years ago, while backpacking through France during a scorching summer, I was impressed by some TV news footage showing thousands of people crammed in the water on the Biarritz coast. It wasn't the mass of humanity that struck me as much as the fact that surfers were catching and riding waves through the throngs. The carnage was unbelievable.

Lifeguards around the world will tell you that among the main causes of injury at the beach are collisions between surfboards, boogie boards, bodysurfers and people. Most surfboards are made of hard fibreglass and have sharp fins. Forget about shark fins, getting impaled or knocked by a board, whether it's yours or someone else's, can be extremely painful. Getting run over and being 'fin chopped', which is a nice term for 'sliced open', is just as bad. Even 'soft' boogie boards can result in a nasty impact.

Many beaches try and keep boardriders away from swimmers by making the flagged swimming areas board-free and using other coloured flags and signs to identify designated surfcraft areas. While surfers should respect these boundaries, so should swimmers. It's also a good idea, no matter where or how you are catching a wave, to always check the water around you and in front of you for loose boards and people before taking off on a wave.

# Beach nasties:
# dangerous marine life

There are two main types of hazards that you need to be aware of at the beach. There are physical hazards, like rip currents

# A quick guide to the Top 5 physical beach hazards

It's important to remember that physical hazards are not only related to natural processes and features that occur on some beaches, but also to the way that we interact with them.

## HAZARD 1 | Rip currents
*Strong and narrow currents that flow offshore*

| RISKS INVOLVED | WHERE AND WHEN | WHAT YOU SHOULD DO |
|---|---|---|
| Quickly carry swimmers seawards | Any beach with breaking waves across a wide area | Relax and stay calm |
| Flow faster than most people can swim | Sandy beaches with breaking waves and sandbars | Stay afloat |
| Difficult to spot to untrained eye | Next to headlands and groynes, jetties and piers | Signal and call for help |
| Often look like safest place to swim | During storms | Swim towards breaking waves and white water |
| Main cause of surf drownings and rescues | | Swim near lifeguards |

## HAZARD 2 | Dumping waves
*Plunging waves that curl over and break with tremendous force*

| RISKS INVOLVED | WHERE AND WHEN | WHAT YOU SHOULD DO |
|---|---|---|
| Strong and violent impact with seabed | Shorebreak of steep beaches | Enter water carefully |
| Head and spinal injuries | Sandbar crests and reefs | Don't surf dumping waves |
| Broken limbs and abrasion injuries | Wherever water depth goes from being deep to shallow over a short distance | Keep arms held out in front of you |
| | | Lay flat and hug bottom as waves pass over |
| | | Don't turn your back on the waves |

## HAZARD 3 | Shallow water

*Shallow bottom and concealed or unseen submerged objects*

| RISKS INVOLVED | WHERE AND WHEN | WHAT YOU SHOULD DO |
|---|---|---|
| Diving headfirst or jumping from great height into shallow water ('tombstoning')<br><br>Head, spinal and other impact injuries | Sandbars, submerged rocks, reefs, structures<br><br>All depths become shallower closer to low tide | Check water depth first<br><br>Enter water slowly<br><br>Arms outstretched when diving<br><br>Don't jump from great heights |

## HAZARD 4 | Surging waves

*Breaking waves that surge up the beach and rush back down rapidly*

| RISKS INVOLVED | WHERE AND WHEN | WHAT YOU SHOULD DO |
|---|---|---|
| Can knock over people, particularly little children and the elderly<br><br>Can drag people back down the beach into the water | On steep and moderately steep beaches<br><br>Often in small wave and protected beaches | Keep eye on waves at all times<br><br>Constantly watch small children<br><br>Brace legs and dig feet in sand |

## HAZARD 5 | Big waves and heavy surf

*Large breaking waves during storms creating extremely energetic surf conditions*

| RISKS INVOLVED | WHERE AND WHEN | WHAT YOU SHOULD DO |
|---|---|---|
| Drowning due to exhaustion from swimming in the presence of large waves, strong currents and whitewater<br><br>Drifting in all directions | Any beach subject to storm waves from strong winds<br><br>Any beach subject to open ocean swell | Look out for the approach of wave sets<br><br>Know your limits<br><br>Don't go in the water when beaches are closed due to dangerous surf<br><br>Stay away from the shoreline to avoid rapid surges of water |

# DON'T RUSH IN

There has been a noticeable increase in recent years of bystander rescuers drowning while attempting to save someone in trouble in the ocean, usually in a rip current. A 'bystander' is any member of the public, be they family, friend or stranger, trying to rescue someone in distress. While many bystander rescues occur and save many lives, particularly by surfers, far too often it is the person trying to make the rescue who ends up drowning, and tragically it's commonly a parent trying to rescue their child.

If you are at the beach and you see someone in trouble, even if it's your child, don't rush in. Fight the urge to sprint to the beach and swim as fast as you can to rescue them. You're already panicking and will likely be exhausted when you reach them, if you manage to get to them at all. It's a recipe for disaster. Instead, you should:

1. Think. Take ten seconds or so to think about what you should do. Those ten seconds are not going to make much of a difference to the person in the water, but could make a huge difference in making sure everyone is going to be okay.
2. Get help. If there are lifeguards or lifesavers on the beach, get someone to go tell them what's going on. If not, get the attention of any surfers who might be in the water or on the beach. If there's no-one, then call for help (if you get reception). But you must seek some sort of help.

3.  Grab something that floats. If you are going to go in the water, look around the beach for a boogie board, a ball, an esky (cooler) – anything that floats. If you bring something that floats, it'll give you and the person something to hold onto. Research backs this up as almost 100 per cent of the bystander rescuer fatalities on beaches in Australia did not bring a flotation device with them.

There are other things to think about – don't go in the water fully clothed and, as hard as it sounds, you need to try and stay as calm as possible. Please don't form 'human chains' to reach someone in trouble in the surf. They look, and are, heroic, but several people at the end of a chain have drowned because they lost their grip and ended up in a rip current. It's also important to realise that the person in trouble is probably not panicking as much as you think they are and in the case of rip currents, there's a good chance that the rip flow will circulate them back into shallow water.

But whatever you do, don't just rush in.

and dumping waves, and biological hazards involving marine creatures that can pack a wallop when it comes to pain and discomfort. While some of the latter are easy to spot, some are almost impossible to see, so it's very important that you understand what critters you could be swimming with.

## Jellyfish

Aside from sunburn and getting stuck in a rip current, the most common hazard you will encounter at an ocean beach is a sting from a jellyfish. Jellyfish are free-swimming animals found in every ocean in the world. These gelatinous blobs come in all sorts of shapes and sizes, and many have tentacles covered in stinging cells that hang down from their body. This is how they capture their food, but unfortunately it's also how people get stung, because the tentacles can be surprisingly long and can easily wrap tendrils around your arms and legs. The severity of jellyfish stings can vary from just a gentle tingle to extreme pain and even death. While this sounds scary, only about 10 per cent of the more than 1000 jellyfish species have stings that will affect humans. Even so, if you see a jellyfish, it pays to keep your distance.

Most jellyfish like warmer water, which is why they are more common in the summer and in the tropics. Probably the best known is the Portuguese man-of-war (known as bluebottles in Australia and New Zealand) which is found in warm oceans around the world. They are easily recognised by a distinctive body sitting on the surface of the water that looks like a blown-up piece of purplish-blue bubble gum (Pic. 35). The body of the bluebottle acts as a sail that allows it to be blown in the direction of the wind. This is why you will often find more of them near the beach after several days of persistent onshore winds. For example, a few days of solid northeasters in the summer on the east coast of Australia usually blows bluebottles in from warmer offshore waters.

The pain from bluebottle stings starts on contact, can be intense and may last for an hour or more. This is often followed by an itchy rash and welts. Unfortunately, if you do get stung, there's not much you can do other than grin and bear it. What you can do before a swim is to check the high tide mark. If there are jellyfish washed up on the beach, chances are there are some in the water as well and it would be wise NOT to go swimming. It's also not a good idea to walk over jellyfish that have been washed up on the beach as their tentacles can still give a nasty sting. Lifeguards will put up warning signs if jellyfish are present. In parts of the United States, a purple flag on the lifeguard tower means that dangerous marine life, usually jellyfish, are in the water.

Much more serious are some of the jellyfish found in the tropical coastal waters of northern Australia and parts of the Indo-Pacific region. Box jellyfish, often referred to as sea wasps or marine stingers, have a smallish bell-shaped body with tentacles up to 3 metres long, but are pale and transparent, making them very difficult to see. Their stinging cells have some of the most potent venom in the world, enough in fact to kill more than 20 adult humans! Maybe it's just me, but this seems a little excessive for something that lives on tiny fish and shrimp.

It's no joke though. Toxins from box jellyfish attack the heart and nervous system and the pain is immediate and so intense that people have been known to go into shock and drown within minutes. For most survivors of a sting, the acute pain can last for weeks and severe scarring can result where the tentacles made contact with the skin (Pic. 36).

# TREATING A
# JELLYFISH STING

Recently I saw a little boy, who had been stung by a bluebottle jellyfish, screaming his head off while his mother laughed that he'd 'be alright' as she rubbed sand into the tentacles still stuck on his leg. This set off even more bloodcurdling shrieks. He didn't look 'alright' to me. That was one traumatised kid who probably won't want to go in the ocean again.

There is a lot of misinformation and confusion out there about how to treat jellyfish stings. If stung, the best advice is to seek professional help immediately, usually from a lifeguard. Recent research suggests that in non-tropical waters, jellyfish stings should be treated with ice unless it's definitely a bluebottle sting, in which case the affected area should be immersed in water as hot as you can stand. You should also try removing any tentacles with your fingers, which have thicker skin and are more resistant to stings.

At best, these treatments will only reduce the pain but not eliminate it. For stings in tropical waters from box and Irukandji jellyfish, the recommended treatment is to pour vinegar on the stings, which kills any remaining stinging cells. In Australia, tropical beaches will have bottles of vinegar available at beach entry points near safety signs. Of course, the best way to treat a sting is to not get stung in the first place, so pay attention to any warning signs and always keep an eye out for jellyfish both in the water and washed up on the beach. Never, never, ever rub sand into the sting ... it's the worst thing you can do.

With a head the size of a thimble, their much smaller cousin, the Irukandji jellyfish, are perhaps more dangerous as their initial stings are subtle – you may just feel an initial sharp pain. The full-blown symptoms may take 30 minutes to appear but are just as severe. If you think you've been stung, seek medical attention immediately.

Marine stingers are more common in the summer months and during the wet season, but even in the winter it pays to talk to the locals or lifeguards before going in the water. If in doubt, don't go out, and if you do, enter the water slowly as some jellyfish will swim away from people given the chance. Fortunately, many popular beaches in stinger-prone areas often have nets in the water, which create safer swimming enclosures (but are not 100 per cent stinger proof!), but it always pays to look for and read any warning signs. It's also a very good idea to wear a full-length lycra body suit in tropical waters. It doesn't only protect you from jellyfish stings, it also protects you from the sun.

## No see-ums: bugs that bite

Years ago, while working on a remote beach north of Yeppoon, Queensland, I was bitten on the bum by a sandfly the size of a pinhead. With the aid of a bit of scratching, tropical heat and generally bad hygiene, this innocent little bite soon turned into a boil (tropical ulcer) the size of a mango, which totally incapacitated me for about a month. Aside from the extreme pain and the embarrassment of having to travel with a doughnut-shaped orthopedic pillow to sit on, I experienced ongoing effects (more boils!) for several more months. The whole story is probably better suited to a medical journal, but to this day I still cannot believe how much discomfort resulted from a tiny sandfly bite.

Although they are not marine animals, biting insects love beaches and they can ruin your holiday if you're not careful. Sandflies, also known as midges or blackflies, are the classic 'no see-ums' because they are so small – you can't see them. They love wet sand in tidal zones and tend to be more common in hotter climates and after periods of rain. Perfect biting conditions are on still, dull and humid days during which sandflies can be voracious, munching all day long. Many people (like me) develop an allergic response that results in extreme itching and large red welts. Scratching, particularly in tropical environments, can often lead to infections, as I found out.

Mosquitoes are a little bigger, a little noisier and just as much of a nuisance, but generally their feeding times are restricted to mornings and evenings and you can often catch them in the act of biting. Unfortunately, in many tropical regions with glorious beaches, some of those mozzies biting you while you are lazing in the hammock may be transmitting serious diseases such as malaria, dengue fever and Japanese encephalitis. Check with a travel doctor about the latest health warnings for countries before choosing an exotic beach holiday location.

The best way to avoid getting bitten by mosquitoes or sandflies is to wear light long-sleeve clothing and generally reduce the amount of exposed skin during biting times. Insect repellents with DEET concentrations of about 30 per cent are also effective and last for about four hours. Dosing up on vitamin B1 about two weeks before holidaying at a bug-prone destination helps promote immunity and reduces allergic response to bites. For Australians, that means putting more Vegemite on your toast and sandwiches!

Biting insects are not just restricted to the air. Sometimes you can feel yourself being bitten in the water and these bites often lead to incredibly itchy welts and rashes on your skin. These are caused by something commonly referred to as 'sea lice'. I always thought I was being 'bitten' by bits of jellyfish tentacles, broken up by the breaking waves, but it would have been sea lice, which are tiny jellyfish larvae found in warm waters around the world.

Sea lice tend to bite areas where they get trapped and squeezed, such as between your swimming costume and your skin and in the armpit area. When sea lice are around, it's often people wearing t-shirts and rash vests in the water who are affected the most. If it felt like you were being nibbled in the water, find a shower as soon as possible after leaving the water, remove your bathing suit and *If it felt like you were being nibbled in the water, find a shower as soon as possible ...* rinse off. Sea lice bites usually have a delayed reaction and can last a long time. Treat them the same way you would severe insect bites and avoid scratching them because this will make them worse.

For years I swam and bodysurfed happily in Australia, mostly at Sydney's Tamarama Beach, without getting bitten by sea lice. When I finally did, I mentioned this to one of the local young surfers, 'Dirty Kev' who said, 'Whatever you do, don't scratch them!' I totally ignored his advice because he was 'Dirty Kev', but I wish I hadn't because they became the itchiest bites I've ever had in my life and it almost drove me mad. I should have listened to Kev because although he presented himself as the archetypal long-haired bleach-blonde surfer and often disgraced himself in public by doing silly things, like vomiting on Jason Donavan's shoes while accepting his

## THE SHARK ATTACK
## SURVIVAL GUIDE | Long-distance ocean swimming
can be extremely pleasurable until you see a strange shadow play across the bottom. It's probably just the reflection of a cloud, but then again, maybe it's … a Great White? Suddenly innocent daydreams are replaced by images of thrashing blood-red water. It can really ruin a good swim. So even though the odds of being bitten by a shark are incredibly small, if it makes you feel better, they can be made even smaller by following some simple rules:

★ Avoid wearing shiny jewellery and bling as they reflect sunlight and can look like small fish to a hungry shark.

★ Avoid swimming and surfing around dawn and dusk (when sharks like to feed and their vision is poor), particularly near murky river mouths (where sharks like to feed and their vision is obscured).

★ Don't swim where birds are diving into the water feeding on large schools of agitated and jumping baitfish that look like they are being pursued by something big.

★ Don't swim alone. Swimming with someone else immediately decreases your odds of being bitten by half!

★ If you are that concerned about sharks, maybe don't go into the water!

If you do see a shark, keep it in sight and move away quietly and calmly, avoiding thrashing motions with your arms and legs. If the shark has targeted you as food and is making a beeline in your direction, one popular theory is to try and punch and gouge the shark in their sensitive snout and eye region. While this is not a proven method, some survivors of shark attacks swear by it.

My own personal theory, based on nothing but a hunch, is that the best way to deal with sharks, crocodiles (and even grizzly bears for that matter) is to turn the tables and attack them first. My thinking is that they're not used to this and will get spooked. I haven't had a chance to test this theory yet, so please let me know how it goes.

second-place trophy in the now defunct Bondi Beach Nude Surfing Competition, he is now a self-made multi-millionaire and owns beachfront property I could only dream of. I always knew he had potential.

## Size does matter: killables in the ocean

A good friend of mine in Toronto has a son named Nick, who when he was five years old told me he would never visit Australia because of all the 'killables' there. I think he was talking about all the snakes and spiders, but I didn't have the heart to tell him that there are plenty more 'killables' in the ocean. And when it comes to the beach, size definitely does matter: it's the little things you have to watch out for. Let's see, there's the deadly blue-ringed octopus, a beautiful little creature the size of a golf ball that inhabits tide pools in the Pacific from Australia to Japan (Pic. 37). Exceptionally shy, it bites only when threatened, but releases a powerful neuromuscular paralysing venom capable of killing a football team. Definitely killable.

Then there are some beautiful and small cylindrical cone shells, highly desired among shell collectors, that can be found on tropical beaches, particularly those near coral reefs. It's not the cone shells that are the problem, but the snails that inhabit them. When handled, some have a tendency to reach out of the shell and fire a tiny poisonous barb, like a harpoon, which contains a paralysing venom that is, of course, killable.

Finally there's the stonefish, possibly the ugliest and deadliest fish in the world – always a bad combination. Stonefish like to lie camouflaged on sandy and muddy seabeds in the tropics, particularly around exposed rocks, and they have dorsal spines sticking out that release venom when stepped on.

This results in some of the most mind-bogglingly excruciating pain a human is capable of experiencing. Extremely killable. For this reason, it's a good idea to wear thick-soled shoes and to shuffle your feet along when walking along shorelines in stonefish habitats.

Before you start cancelling your vacation plans to the tropics, it's important to remember that encounters with any of these animals are extremely rare. The best advice for dealing with all these deadly little things is: don't get bitten in the first place. If you understand where they live and use common sense, these shy little animals that just want to be left alone will remain harmless. A good rule to follow when it comes to any sea creatures on the beach, or in any marine environment, is always: LOOK, BUT DON'T TOUCH.

## Shark!

I have spent years trying to convince my friend Pierre that he has nothing to worry about when it comes to sharks, and felt that I had made some real progress, but when I told him I was putting his fear of sharks into this book, all he said was, 'Sharks kill man … that's the bottom line'. Sigh. But do they? If you must insist on placing sharks at the top of your 'Things that Terrify You About the Ocean' list, then you should also know that your odds of getting bitten by a shark are infinitesimally small and if you are unlucky enough to be bitten, the chances of dying from the bite are less than 10 per cent. Putting this in perspective, you are 30 times more likely to be hit by lightning at the beach. Don't laugh, I got zapped a few years ago, but I still haven't seen a shark.

However, shark bites are no laughing matter and beyond the statistics, the emotional and economic effects on the people involved and the associated communities can be significant. In Australia, the number of shark bites per year has risen from an average of nine per year from 1990 to 2000 to 22 bites per year from 2010 to 2020 and over the last ten years about two to three people per year have died as a result of their injuries. Globally, there are about 80 unprovoked shark bites and eight fatalities per year. Yes, there's evidence that the number of shark incidents is increasing, but so is the number of people going in the water and interacting with the ocean in different ways so it's really hard to say whether anything has really changed. A fatal shark attack at a Sydney beach in 2022 was the first in the city in 60 years. The odds of not being bitten are still very much in your favour.

I don't want to be bitten by a shark and I feel terrible for anyone who has and the families and friends who have been affected, but the vast majority of shark bite survivors hold no ill will towards sharks. The ocean is the shark's domain and to be honest, it's a risk you take when you enter the ocean. In my opinion, far too much funding is devoted to the 'shark problem' when it could be used to reduce the drowning numbers on our beaches in many other ways. For example, there's a lot of money being invested into drones to spot sharks when they'd be much better being used to map the location of rip currents, but I'm biased ... I know that you are about 20 more times likely to drown in a rip current than be killed by a shark! However, I totally respect shark fear and here is one last suggestion: use the displacement-of-anxiety approach and worry about something else instead, like whether or not someone is going to take your phone that you left lying on your towel on the beach. This will help keep your mind off the sharks.

# THE BEACH SAFETY CHECKLIST

## BASIC STUFF

✓ Be sun smart and Slip, Slop, Slap, Seek and Slide before heading to the beach and keep it up when you're there.

✓ Drink lots of water to avoid dehydration and sunstroke.

✓ Check the internet for weather, tide and surf conditions. Better yet, get some decent apps that do this for you like the Surf Life Saving Australia Beachsafe app, which also tells you what beaches are closest to you and whether or not they are patrolled by lifeguards and lifesavers.

✓ Make sure someone knows where you're going.

✓ Take only what you need and leave your valuables at home. Unfortunately, theft on busy beaches can be a problem.

## AT A PATROLLED BEACH

✓ Always STOP and spend a few minutes THINKING about beach safety. LOOK for hazards and have a PLAN if something goes wrong.

✓ Pay attention to any beach flags, read any warning signs and always swim in a lifeguarded area. If they can't see you, they can't save you.

✓ In countries where it applies, like Australia and New Zealand, ALWAYS SWIM BETWEEN THE RED AND YELLOW FLAGS. It's a safer area supervised by lifeguards and lifesavers. In other countries, make sure you understand what the beach flag colour means in terms of hazard levels.

✓ Ask the lifeguards about the surf conditions and where it is safest to swim. Ask them to point out any rip currents to you.

✓ Make sure that all young children are accompanied by an experienced adult swimmer.

## AT AN UNPATROLLED BEACH

✓ Always STOP and spend a few minutes THINKING about beach safety. LOOK for hazards and have a PLAN if something goes wrong.

✓ If you can't swim or are an inexperienced ocean swimmer, don't go in. If you do, make sure to not go in past waist depth and always make sure your feet are firmly on the bottom.

✓ Regardless of your experience, know your limits. Never go out in conditions you are not suited for. If in doubt, don't go out.

✓ Look for rip currents and think about whether the wave conditions are too rough for you.

✓ Make sure someone knows where you are and make sure you have a phone with you for emergencies.

✓ Bring something that floats with you to the beach, like a boogie board or ball. If not, when you get there, look on the beach for something that floats. If someone gets in trouble, this will help them (and you – if you try to rescue them) to stay afloat.

✓ If you find yourself in trouble in the water, stay calm, float and signal or call for help. There might be surfers nearby.

✓ If you are the only person on the beach, it's probably not a good idea to go in!

# The bottom line

★ Always slip on a shirt, slop on some sunscreen, slap on a hat, seek some shade and slide on some sunglasses when you go to the beach. Drinking lots of water is also a good idea.

★ Try to choose a beach with lifeguards or lifesavers and swim between the red and yellow flags in Australia and New Zealand.

★ Read and pay attention to any beach safety signage.

★ Watch out for other swimmers, surfers and boardcraft in the water.

★ If you see someone in trouble in the water, first call for help; if you do go in to help them, always bring something that floats with you.

★ Rip currents are the main hazard you'll find on surf beaches, make sure you know what they are and how to spot them.

★ Be careful of dangerous breaking waves at the shoreline (shorebreak).

★ When it comes to marine animals, look but don't touch.

★ Don't worry about sharks, the odds are very much in your favour that you'll never even see one.

★ Always spend a few minutes thinking about beach safety when you go to the beach. Look for any hazards and have a plan if something goes wrong.

# Acknowledgments

There wasn't a lot of surfing and beach time when I was growing up on the shores of Lake Ontario in the suburbs of Toronto, Canada, particularly not with a nuclear power plant down the road. Most of what I know about beaches came from travelling, experience and studying, but most of all from other people. I've been told that this section was rather long for the first edition, but I'm always amazed how short most acknowledgment sections are. I've got a lot of people to thank – even more thirteen years later! I apologise in advance for anyone I've missed, but in the world of coastal geomorphology and beach safety and in geographical and chronological order I'd like to thank Brian Greenwood, Phil Osborne, Bernie Bauer, Doug Sherman, Troels Aagaard, Rowland Atkins, Andy Short, Peter Cowell, Gerd Masselink, Ian Turner, Gui Lessa, Patrick Hesp, Michael Hughes, Justin Meleo, Thomas Williams, Peter Nielsen, Bruce Thom, David Edwards, Paul Kench, Roger McLean, Jamie MacMahan, Wendy Carey, Spencer Rogers, Greg Dusek, Stephen Leatherman, Patrick Rynne, Chris Brewster, Bruno Castelle, Tim Scott, Jak McCarroll, Chris Houser, Caroline Finch, Ann Williamson, Julie Hatfield, Shauna Sherker, John Andrews, Adam Weir, Shane Daw, Will Koon and all the lifeguards and beach safety types I've collaborated with over the years. I have such unbelievable respect for professional beach lifeguards.

The book wouldn't be possible without the help of the Tamarama Beach Surf Life Saving Club, who let me be their caretaker for four years, allowing me to write my PhD in my 'budgie smugglers' while living next to and learning from one of the best beaches on the planet. I also learned a lot about reading the surf from many of my fellow lifesavers, who took a 'seppo' under their wing, and from the stellar Waverley Council lifeguards over the years, particularly Brendan Reade, Bill Moore and Bruce Hopkins.

'The Science of the Surf' (SOS) certainly wouldn't have been as successful as it is without the early support of Phil Hogan, Jon Hancock and all the Tamarama SLSC members who helped chuck dye in the rip and did not drown. Also deserving thanks for support are Chris Tola, the Sydney Coastal Councils Group, numerous local government councils in New South Wales, particularly Randwick City Council, for supporting my community presentations, Surf Life Saving Australia (SLSA), the Australian professional Ocean Lifeguard Association (APOLA), the NSW Department of Education and Training, and all the primary and high schools who have hosted SOS. The University of New South Wales (UNSW Sydney) has been incredibly supportive of my beach safety work, particularly the Faculty of Science and my Heads of School (Paul Adam, David Cohen and Alistair Poore). And Mary O'Malley and UNSW TV were instrumental in getting what was one of the first educational rip current videos out on YouTube – 'How to Survive Beach Rip Currents' – which now has 1.6 million views and counting. Jason Markland and Steve Kudzius were kind enough to include my work in some amazing full-length documentaries about rip currents that are incredibly powerful and educational and are no doubt saving lives. If you don't believe me, follow the link to 'Rip Current

Heroes' on <www.scienceofthesurf.com> and also check out <www.ripcurrentsafety.com>.

Thanks to NewSouth Publishing, particularly Harriet McInerney, for continuing to believe in this book and encouraging me for years to do a second edition. My editor Paul O'Beirne made the transition to this new edition incredibly smooth, for which I am extremely grateful. There are many excellent coastal textbooks and books about beaches out there, but I refused to look at them. Any similarities to other works are due to the generic nature of beach science, the scattered remnants of my memory and coincidence. Finally, a huge thank you to my family – my wife Danielle and my daughters Layla and Ivy – who somehow continue to put up with me, but through osmosis have learned how to spot rips.

# Index